The Cultured Chimpanzee
Reflections on Cultural Primatology

Short of inventing a time machine, we will never see our extinct forebears in action and so be able to determine directly how human behaviour and culture have developed. However, we can learn from our closest living relatives – the African great apes. *The Cultured Chimpanzee* explores the astonishing variation in chimpanzee behaviour across their range which cannot be explained by individual learning, genetic, or environmental influences. It promotes the view that this rich diversity in social life and material culture reflects social learning of traditions, and more closely resembles cultural variety in humans than the simpler behaviour of other animal species. This stimulating book shows that the field of cultural primatology may therefore help us to reconstruct the cultural evolution of *Homo sapiens* from earlier forms, and that it is essential for anthropologists, archaeologists, and zoologists to work together to develop a stronger understanding of human and primate cultural evolution.

WILLIAM C. MCGREW is Professor of Anthropology and Zoology at Miami University in Ohio. He has studied the socioecology of wild chimpanzees throughout their range – from Senegal to Tanzania – for over 30 years. Among other works, he has written *Chimpanzee Material Culture. Implications for Human Evolution* (Cambridge, 1992) and edited *Great Ape Societies* (Cambridge, 1996) with Linda Marchant and Toshisada Nishida.

The Cultured Chimpanzee

Reflections on Cultural Primatology

W. C. McGREW
Professor of Anthropology and Zoology
Miami University (Ohio)

CAMBRIDGE
UNIVERSITY PRESS

PUBLISHED BY THE PRESS SYNDICATE OF THE UNIVERSITY OF CAMBRIDGE
The Pitt Building, Trumpington Street, Cambridge, United Kingdom

CAMBRIDGE UNIVERSITY PRESS
The Edinburgh Building, Cambridge, CB2 2RU, UK
40 West 20th Street, New York, NY 10011–4211, USA
477 Williamstown Road, Port Melbourne, VIC 3207, Australia
Ruiz de Alarcón 13, 28014 Madrid, Spain
Dock House, The Waterfront, Cape Town 8001, South Africa

http://www.cambridge.org

First published 2004

Printed in the United Kingdom at the University Press, Cambridge

Typefaces Trump Mediaeval 9.5/15 pt. and Times   *System* LaTeX 2_ε [TB]

A catalogue record for this book is available from the British Library

Library of Congress Cataloguing in Publication data
McGrew, W. C. (William Clement), 1944–
The cultured chimpanzee : reflections on cultural primatology/W. C. McGrew.
  p. cm.
Includes bibliographical references (p.) and indexes.
ISBN 0 521 82841 4 – ISBN 0 521 53543 3 (paperback)
1. Chimpanzees – Behavior. 2. Animal societies. 3. Psychology, Comparative.
I. Title.
QL737.P96M442 2004
156 – dc22   2004045828

ISBN 0 521 82841 4 hardback
ISBN 0 521 53543 3 paperback

'The animals themselves are always more important than the books that have been written about them.'

Niko Tinbergen (1953)

'Chimpanzees are always new to me.'

Toshisada Nishida (1993)

Contents

Preface

One of the inspirations for this book is another slim volume, Jane Lancaster's (1975) *Primate Behavior and the Emergence of Human Culture*. Although written almost 30 years ago, it foreshadowed many of the issues taken up here, especially the implications of ape behaviour for modelling the evolutionary origins of humanity's complexity. There is a prophetic chapter, 'Social Traditions and the Emergence of Culture'. She worked with what was known at the time, especially the innovative food-processing of Koshima's Japanese monkeys and the elementary technology of Gombe's chimpanzees (see below). Much of what she said holds today, but much has changed. For example, there was no hint in her book of comparative analysis of cultural variation across primate communities. Further, all of the examples described for chimpanzees were for material culture (although there was a pioneering treatment of play in vervet monkeys), upon which I focused in an earlier book (McGrew, 1992). Absent, because they had not yet been studied, were the nonmaterial cultural aspects of social relations and structure, and communication. These are treated here.

In the earlier book, I tried to set out my biases, and disappointingly, they all remain: naturalist (not experimentalist), empiricist (not theoretician), publisher (not story-teller), monolinguist (not polyglot), and evolutionist (not creationist). In the last 10 years, as cultural primatology has emerged as an entity, it has done so in parallel with the 'culture wars' in the arts, humanities, and even the social sciences. Perhaps only the word *ecology* has been more misused than the word *culture* in this period. This state of affairs has made it an interesting time to try to keep one foot in social science and the other in natural science. One of the results of the strife (at least for me) has been the affirming realisation that the most important single base for this book and the work it contains is the scientific method. I am first, last, and always, a scientist, and all the discourses, texts, and voices in the world are not worth a jot, by comparison with a good hypothesis tested with clear data. If that is not a nailing of colours to the mast, what is? If it damns me in the eyes of contemporary cultural anthropologists, so be it.

Meanwhile, I shall fly today to Conakry, hoping to see the chimpanzees of Guinea engage in the lithic technology of nut-cracking. Theoretical disputes are the last thing on my mind, by comparison with the prospect of seeing apes in action.

Oxford, Ohio, 1 August 2003

Acknowledgements

This book was wholly written while I was a visiting research fellow at the Leverhulme Centre for Human Evolutionary Studies, in the Department of Biological Anthropology, University of Cambridge, all of whom I thank for their support. I am grateful especially to Rob Foley, Marta Lahr, Mike Petraglia, and Jay Stock for their intellectual companionship. At the same time, I was a visiting bye fellow at Selwyn College, Cambridge, and I thank them for their hospitality, especially David Chivers and Ian Thompson.

For help in over 30 years of chasing chimpanzees, there are too many persons to be thanked here by name. Most of my field companions from 1972–91 are mentioned in the preface of my *Chimpanzee Material Culture* (1992). Since 1992, I am grateful for collegiality in the bush to: Anthony Collins, Craig Stanford, Charlotte Uhlenbroek, Bill Wallauer and Janette Wallis at Gombe; Moshi Bunengwa, Mike Huffman, Kenji Kawanaka, Michio Nakamura and Shigeo Uehara at Mahale; Adam Arcadi and Richard Wrangham at Kanyawara; Jill Pruetz and Peter Stirling at Assirik; Mboule Camara, Philip Fulton, Susannah Johnson-Fulton and Djanny Kanté at Fongoli; Tanya Humle and Yukimaru Sugiyama at Bossou. Special thanks go to Linda Marchant, who was there with me at all of these sites, except Fongoli. Without these good colleagues, little could have been accomplished.

Field research abroad requires funding, and for this I am most grateful to the Rebecca Jeanne Andrew Memorial Award (Miami University), the Philip and Elaina Hampton Fund for Faculty International Initiatives (Miami University), the L. S. B. Leakey Foundation, National Geographic Society, the National Science

Foundation (BSC-0122518), Max Planck Gesellschaft, and Primate Conservation Inc.

Each of us has experienced the frustration of writing about a topic, only to discover after it was published that someone else was writing and publishing the same ideas in parallel. This messiness in science is understandable and is part of the game, but it can be reduced by sharing unpublished material, whether this be manuscript, datum, or photograph. I have been blessed by the generosity of the following colleagues, many of whom did not agree with my point of view but who did such sharing with me: Michael Alvard, Christopher Boehm, Christophe Boesch, Gillian Brown, Richard Byrne, Robert Foley, Dorothy Fragaszy, Michael Huffman, Marta Lahr, Kevin Laland, Louis Lefebvre, Linda Marchant, Michio Nakamura, Toshisada Nishida, Melissa Panger, Elizabeth Pimley, Luke Rendell, Zhanna Reznikova, Carel van Schaik, Sabine Tebbich, Ifke van Bergen, Hal Whitehead, Andrew Whiten, and Richard Wrangham. If I have forgotten anyone, please tell me, and I will buy you a beer.

This book would never have been written, were it not for Tracey Sanderson, the commissioning editor at Cambridge University Press, who shepherded it all the way from lunch in Buffalo to completion. She scrutinised it chapter by chapter during those 6 months in Cambridge, and our periodic editorial meetings were essential to keep me on track. I am truly grateful, and insist on this too-short paragraph.

I still write in longhand, so someone else had to word-process the text, and for that I am most grateful to Diana Deaton and Carol Kist, who saw so many versions of it sent back and forth trans-Atlantically that they will be glad to see its backside. Lauren Sarringhaus helped me put the first draft into final form, Alysha Kocher, Daniel Pesek, Jacklyn Ramsey, and Samantha Russak helped with the copy-editing and the indexing. Cara Wall Scheffler found the back-cover image of the human version of the grooming hand-clasp, and most importantly, Linda Marchant made critical comments on the whole text. In the end, of course, all remaining errors are my responsibility.

Finally, my deepest gratitude goes to two sets of beings whose lives are intertwined: the apes and the people who keep them safe. It is likely that the chimpanzees never knew or cared that I was there to study them, but at least some of them tolerated my presence, and so made my life easier. I admire even the ones who fled from me (however much I cursed them at the time), for their dignity and tenacity in a world that seems hellbent on extinguishing them. To the people in conservation, from those who sit in air-conditioned offices in capital cities to the guards, rangers, and field assistants in the bush, words cannot express my respect for your dedication, in the face of being overworked and underpaid, year after year, and now decade after decade. The chimpanzee might have survived the last century without your protection, but certainly will not in the current one. You are both truly reasons for hope.

I Introduction

My first inkling of cultural primatology came in January, 1975, when Caroline Tutin and I went to the Mahale Mountains of western Tanzania. At that point, I had 15 months of field experience studying the wild chimpanzees of Gombe and thought I knew the species. On the first day out at Kasoje, this was proven wrong. We saw the chimpanzees of K-group doing the grooming hand-clasp, something neither we nor anyone else had ever seen at Gombe, in thousands of hours of observation (see Figure 1.1). We were dumbfounded by its elegant symmetry. However, upon returning to camp, when we mentioned the discovery to our host, Professor Junichiro Itani, he was unimpressed. Did not all chimpanzees do this?

At that point, I realised that there was no such creature as The Chimpanzee, if by that was meant behavioural uniformity across the species *Pan troglodytes*. Instead, there was behavioural diversity across chimpanzees, apparently at the level of populations. So, how to explain this variation? This book describes my attempts to answer that question, and many other related ones, over the last 25 years.

As an ethologist, I knew about within-species variation in animal behaviour. One of my fellow postgraduate students at Oxford, Michael Norton-Griffiths, had done elegant cross-fostering experiments on oystercatchers, showing that parental foraging and food-processing techniques were passed on from parent to offspring (Norton-Griffiths, 1967). But that was diet, and diet is influenced readily by the availability, abundance, or distribution of food items. In contrast, the grooming hand-clasp at Mahale was a social behavioural pattern, apparently arbitrary in form and independent of any obvious environmental constraints; it is essentially a mutual gesture.

FIGURE 1.1 Adult female (left) and adult male of K group, Mahale, engage in grooming hand-clasp. Note infant in her lap, asleep but still clasping hair of his mother's side.

As a primatologist, I had admired the pioneering studies of Japanese field workers, studying the Japanese monkeys, especially on Koshima (Kawamura, 1959). Individuals invented new techniques of food processing or of thermoregulation, and these were taken up by fellow troop members. The scientists termed this 'protoculture' (Itani & Nishimura, 1973). But each of the behavioural patterns, whether sweet-potato washing or hot-spring soaking, was instigated by humans; it was not spontaneously shown by the monkeys (Matsuzawa, 2003). In contrast, the grooming hand-clasp of the Mahale chimpanzees had nothing to do with humans, being instead an expression of ape sociability.

Finally, as a comparative psychologist, I knew of the hefty published literature on song-learning in passerine birds. Complementary studies in field and laboratory of its nature and nurture had established that traditions of vocal learning exist, passed on from generation to generation (Slater, 1986). In some cases this led to regional dialects, and sometimes these vocal variations were shaped by interaction with others, even with non-singers (West et al., 2003). But these same songbirds showed no other impressive feats of social learning; they were essentially one-trick ponies with a very good trick. The hand-clasping chimpanzees, on the other hand, showed many other examples of behavioural diversity, from tool use to courtship.

These, and all the other explanations for phenotypic variation that were available then from the natural sciences, did not seem enough to explain what the chimpanzees were doing. We needed to look further afield, and the obvious alternative was the social sciences. Enthusiastic but naïve, we wrote up our Mahale findings and submitted them to a prestigious journal, Man, the Journal of the Royal Anthropological Institute of Great Britain and Ireland. Thanks to the intellectual generosity of its editor, Peter Loizos, the article was published (McGrew & Tutin, 1978). Apart from one damning response (Washburn & Benedict, 1979 (see below)), our discovery sank without trace. The late 1970s was not a time to champion cultural primatology, at least in the West.

Table 1.1 *Levels of analysis for cross-cultural comparisons of chimpanzees. Note hierarchical embeddedness descending from left to right.*

Level	Example
Species	*Pan troglodytes* versus *P. paniscus*
Subspecies	*P. t. troglodytes* versus *P. t. schweinfurthii* versus *P. t. verus*
Population	Mahale (Tanzania) versus Kibale (Uganda)
Community	Kanyawara versus Ngogo communities at (Kibale)
Clan	'F' versus 'G' lineages at (Gombe)
Individual	Frodo versus Freud, in 'F' lineage

My overall point is a simple one: faced with a dataset of ape behavioural diversity that demanded explanation, we were forced to invoke *both* the natural and social sciences for help. Thus, we had to try to straddle an intellectual divide in which one side believed that evolutionary theory explained everything, and the other that it explained nothing. Even now, 25 years later, biocultural anthropologists still catch flak from both sides of the divide (Cronk, 1999).

LEVELS OF STUDY

Consider behavioural diversity (see Table 1.1). To explain it requires clarity about which levels of analysis are being compared, and lack of this can lead to confusion, or worse. Comparison of apples with oranges yields a confusing fruit salad, so it is worth setting out this embedded structure now, as it will come up again. The top and bottom levels of the six-level schema in Table 1.1 are the easiest to tackle.

Comparing closely related *species* such as the chimpanzee and the bonobo is necessary to reconstruct their phylogeny or to infer the behavioural repertoire of a common ancestor. But differences between species are not likely to be cultural, since by definition they have different genotypes that could account for the variation, and occupy different ecological niches.

At the other extreme, comparing *individuals* within a species can be revealing, especially with regard to issues of ontogeny or personality. Much of psychology depends on this level of comparison. For chimpanzees at Gombe, the fact that, of the two brothers, Frodo acts like a bully and Freud appears relaxed in outlook, is fascinating, but one would be most likely to ascribe these individual differences to upbringing or character, not to culture.

The four intermediate levels each present a different set of issues with regard to explaining behavioural diversity in some kind of cultural terms as follows.

(1) At the *subspecies* level, chimpanzees are divided into three recognised geographical races across Africa: central, eastern and western. There is a purported fourth subspecies, found between the central and the west, in Cameroon and Nigeria, but it was described only in genotypic terms (Gonder *et al.*, 1997) and is only now being studied in nature (Sommer *et al.*, 2003). Comparison on the broadest, species-wide scale can be done at the subspecific level, from the mountains of Tanzania, to the lowland forests of the Congo basin, to the savannas of Senegal. But the three-way split is not really equal: the western chimpanzees differ genetically from the others far more than eastern and central differ between themselves (Morin *et al.*, 1994). These subspecies are real, in that gene flow between them is now prevented by zoogeographical barriers, especially rivers. Only the western chimpanzees use percussive technology to crack open nuts, and the boundary between crackers and non-crackers seems to be the Sassandra-N'Zo river in Ivory Coast (Marchesi *et al.*, 1995).

(2) Within a subspecies are its constituent *populations*, which amount to demes. These were probably larger in area and numbers until recently, when human activities, especially deforestation, have fragmented them. Not that long ago, all of western Uganda was probably a single megapopulation of chimpanzees; now it is a string

of beads of various-sized blocks of forest. Most of the analyses invoked in this book are at the level of population, making use of the well-known study sites of Gombe, Mahale, Taï, Bossou, etc. (A guide to study sites of chimpanzees is given in Chapter 6.) If there were to be a Chimpanzee Relations Area File to match the Human Relations Area File (that venerable repository of ethnographic data) then it would be at this level.

(3) Within a population are the actual *social units* of which individuals are members. For chimpanzees, these are called variously 'communities' or (unit)groups. These are not troops of constant association – as in many monkeys – nor are they families – as in lesser apes – but units that rarely (if ever) are all together in one place at one time. Instead, they fuse or fission into parties of varying composition: all-male parties for patrolling, nursery aggregations of mothers and dependent offspring, consorting mating pairs, etc. Yet the members are bound up in a social unit, whether it is males co-operating in territorial defence or females emigrating to breed. In one case, the population is only one community (Bossou), in another only two communities probably survive (Gombe), and in some populations, there are enough communities that we cannot be sure of their number (Mahale). Unfortunately, for the sake of comparison, little is published on neighbouring groups, or what is known is uneven or not coincident in time. Whenever possible in this book, comparisons will be drawn at this cultural level, e.g. K-group versus M-group at Mahale, and Kanyawara community versus Ngogo community at Kibale.

(4) Finally, there are kin-based subdivisions within the unit group. These *clans* seem to be matrilineal, but our knowledge of patrilineality was constrained until recently by lack of knowledge of paternity (Morin *et al.*, 1994; Vigilant, 2002). Almost nothing is known of chimpanzee cultural life at this level, though speculation has been provocative, e.g. Goodall (1986b) told of the F-lineage's style of mothering, passed down through three generations from Flo to Fifi to Fanni at Gombe. Clan influences clearly are there to be

investigated: the alpha male position at Gombe has been held for most of the last 20 years by members of either the F or G lineages. By the time this appears in print, the balance may well have tipped again, with Gimble ousting Frodo.

If apes were humans, any of the four levels of subspecies, population, community, or lineage would be fair game for study by social scientists, especially sociocultural anthropologists.

HUMAN UNIQUENESS

But apes are not humans. That simple fact could stop cultural primatology in its tracks. If only humans are cultural, then social scientists need look no further than our species for subject matter. To address this point requires a consensual definition of the phenomenon, which is a subject for the next chapter. Assume for the moment that culture somehow can be defined in a way that encompasses what humans do and are, but also can be applied operationally to other species. Even if this possibility is granted in principle, there are several strong objections to considering it in practice.

One objection is the obvious point that *Homo sapiens* is unique. This is true, but by definition so is every other species (Foley, 1987). Species uniqueness does not prevent us from using the comparative method to undertake (for example) immunology, or to elucidate a phenomenon like lactation, in our closest living relations. A blood transfusion from a human could save a chimpanzee's life, or vice versa, but only with the right combination of A–B–O blood types. When we test the efficacy of vaccines or the effects of drugs on cognitive performance in chimpanzees, we do so for the very reason that we and they are fellow hominoids sharing many key traits.

But surely (some say) humans are behaviourally, cognitively, and emotionally unique, and that is what determines whether or not a creature is cultural. We are the only species known to pilot aircraft, invent calculus, and celebrate St Valentine's Day. Fair enough, but most humans, either as individuals or as societies, do none of these.

That makes them no less human, any more than prelinguistic infants, nonlinguistic autists, or postlinguistic elders are denied their humanity because they fail the language competence test.

The true measure of human species uniqueness is to know which (if any) of its *universal* traits are *qualitatively* different (in kind) from the universal traits of chimpanzees. Quantitative differences (in degree) are not enough. We do not deny human status to societies that have the abacus but not the computer, or that treat illness and injuries but lack the germ theory of disease. Examination of a few candidate traits will show the difficulties in pinpointing human cultural uniqueness.

Bipedal locomotion is a form of behaviour that is universal to humans; we all walk and run (Hunt, 1994). At first glance, upright bipedality also seems unique among primates, too, as great apes are habitual quadrupeds. But a second glance shows that apes can assume bipedal stance or gaits if the context is apt (Hunt, 1992), and in exceptional cases, a wild ape can become even habitually bipedal (Bauer, 1977). A black-and-white difference starts to look grey.

Or take mathematical ability as one aspect of cognition of which humans justifiably are proud. Even the simplest foraging societies have some system of numbers, even in nonliteracy. No one has yet seen a wild chimpanzee count on her fingers or sort sets of items by number. If this happened, we would be entitled to infer numeracy. Meanwhile, studies of captive chimpanzees put into situations where numerical ability yields a payoff show that they can do arithmetic (Boysen & Hallberg, 2000; Biro & Matsuzawa, 2001). More recently, American coots in nature have shown the ability to count their eggs in the nest, to guard against brood parasites who slip in their alien eggs (Lyon, 2003). Thus, an apparent qualitative difference now looks more quantitative.

As for emotions, we can but infer them on the basis of behaviour, whether by vocalisation, facial expression, gesture or posture, in terms of what resonates with us by similarity. Tickle a chimpanzee, and she laughs; startle a chimpanzee, and he grimaces; threaten a chimpanzee, and she lashes out; groom a chimpanzee, and he sprawls relaxed. All of these signals of feelings are recognised readily by the average person. More dramatically, when we see an orphaned ape with her dead mother, her demeanor or 'body language' is one that, if seen in a human child, would be interpreted as grief.

So, are there any qualitatively unique traits shown by humans that could be used to deny culture to other species? The determined sceptic will fix on language. No one has ever heard a wild ape speak. In practice, it can be said that what we know about one another as humans comes from verbal report. You talk and I listen. Most of socio-cultural anthropology, and therefore most of what we know about human culture, is based on the speech of native informants. Once translated, transcribed, deconstructed, etc., the recorded texts reveal the meanings that underlie and permeate the human condition.

What nonsense! Speech is behaviour too, just like another observable action. Words are voluntary puffs of air, and so they need not reflect reality in the slightest. Large-brained, intelligent creatures practice deception, and one of the easiest ways of doing so is by telling lies. (Ask yourself honestly, when the truth matters, do you pay more attention to the words or to the accompanying non-verbal signals?) Why should an anthropologist believe an informant's words to be true, any more so than any other human being seeking to learn something from a companion in everyday life? Yet the corpus of ethnography, at least with regard to ascription of meaning, is based on this tenuous premise. The situation actually is even more vexed: given a creature intelligent enough to be prone to self-deception, then even the most honestly intended verbal report may be false. We do not know if apes are self-deceivers, but humans are.

So, is language a methodological curse or blessing for cultural anthropology? Clearly, if words are predictive of action, then like any other signal, they can be valid and reliable data. ('When we paint our faces blue, we are ready to go into battle.') On the other hand, words that refer to mental or emotional states never can be verified. ('Because I dreamt last night of battle, today I will paint my face blue.') An extreme view is that language is useful in studying culture only when it is corroborated by other acts. ('Back off, or I'll smash your blue face in!') A less extreme view is that language alerts us to attend more to some things than to others. ('We only paint our faces blue when the moon is full.') But many verbal statements will always be obscure ('We paint our faces blue when the gods of war are angry.')

One possible insight into the problem would be to videotape people without their knowledge, thus guaranteeing a record of their spontaneous behaviour. Then, two sets of naïve viewers would scrutinise these 'candid camera' tapes, seeking to understand what was going on. One set would have the sound turned on, the other off. To what extent would the latter data be impoverished? By 10, 50, 90 per cent? So far as I know, the experiment has never been done, but the videotaped material is available (Eibl-Eibesfeldt, 1989).

Are observers of non-linguistic chimpanzee behaviour in the same position as video viewers with the sound turned off? Yes, in many ways, which means that their data are sparse but probably valid. No cultural primatologist will secure an interview with a wild ape, but by the same token, no observer of apes will get lied to. (Actually, it is not quite so pat. As apes have been shown to practice nonverbal deception among themselves (Byrne & Whiten, 1988), they also may do so to human researchers.)

The upshot of all of this is that cultural primatologists studying nonlinguistic apes and cultural anthropologists studying linguistic humans both have costs and benefits with which to cope. In the end, both draw inferences, and the usefulness of the resulting knowledge is a function of how good they are at inferring. In either case, they can do ethnography, even if their methods differ. Sometimes, the results will readily be comparable, and so the accuracy of inferences will be high, e.g. if both humans and apes use stones as hammers to crack nuts, then we compare not only their artefacts, but also their ways of acquiring and using them to solve a simple problem: how to extract a kernel from a shell?

Other times, the inference will be tricky. Suppose humans and apes live in the same forest, and both decline to eat blue duiker? We ask the humans 'Why?', and they tell us that consumption of the species is tabooed. This clearly seems cultural. We then observe the apes and see that they are skilled hunters of other duikers, which they eat avidly. It could be that chimpanzees also taboo blue duikers, but how can we ever know for sure? Or it could be that blue duikers are so wily that neither species even bothers to try to hunt them!

Speaking of taboos, which of the following have something in common: custom, tradition, rite, convention, ceremony, institution, role, style, initiation, etc.? All are nouns referring to collective phenomena that characterise human culture. No one would deny their ubiquity in *H. sapiens*, from the simplest foraging culture to the most complex industrial one. Yet anyone who tries to apply such a term to nonhuman sociality risks derision, if not rejection, at having committed the sin of anthropomorphism. How could a group of apes perform a ceremony? Or subscribe to an institution? Or employ a style? The whole idea is ridiculous, for such phenomena require consciousness, intentionality, imagination, and ability to use symbols and attribute meanings. Everyone knows that only humans can do this.

There are two obvious responses to such arrogance. One is that anyone following the experimental cognitive sciences as applied to other animals in the last 20 years will know that every one of the above capacities has been claimed in some form for nonhuman species. Whether whale or dolphin, parrot or crow, monkey or ape, there is a burgeoning scientific literature on complex comparative cognition (e.g. Tomasello & Call, 1997; Whiten, 2000).

The second response is more basic. Short of mind-reading, one cannot divine whether the behaviour of an ape is a rite or not, because a rite is an idea, not an act. For example, the Christian rite of Communion is more than swallowing a wafer and some wine. To be precise, we (the observers) *think* that it is more than mere ingestion, because its practitioners *say* that it is. Listening to the words of the ceremony, we are *told* that they believe that flesh and blood are being consumed. But, what if we watch and listen to aboriginal converts, worshipping with missionaries upon whom their health care depends: what are we to infer about their real belief and faith? We might suspect that they would utter the words regardless of their truth status, since their lives might depend on their perceived piety.

So, we are back to the pros and cons of relying on verbal report, if we propose to study a nonlinguistic species. If we cannot interrogate an ape, then we must rely on observable acts and be agnostic about what underlies them. It is entirely possible that apes in nature have all of these symbol-laden phenomena. It is also possible that they do not. One must recall that old saw: 'Absence of evidence is not evidence of absence.' Studying culture in any species is not easy, but cultural primatologists and cultural anthropologists have enough in common that they can help one another.

PALAEOCULTURE

Despite all of the above, it may be that the cultural gap between living humans and apes is so great that they cannot be understood in a common framework. Does this get cultural anthropology off the hook of considering other species? No, because we shall have to explain the culture of extinct, intermediate forms in the human lineage, and its offshoots (cf. Foley, 2001, 2003). We have to explain the cultures of the Neolithic, Mesolithic, and the Paleolithic, all the way back to the Last Common Ancestor (LCA) of living humans and apes, at least 6–7 million years ago. Furthermore, recent archaeological findings on the lithic technology of wild chimpanzees suggest that we have to explain the cultural heritage of ancestral apes as well (Mercader *et al.*, 2002; Foley & Lahr, 2003).

If drawing inferences about living, behaving, creatures is hard, then doing so for long-dead ones is hellish. All the behaviour is long gone, though sometimes its products remain, e.g. the fossil footprints of Laetoli, dated at 3.5 million years ago. However rare, such products can be telling, for example, the Laetoli footprints are the oldest undeniable evidence of upright bipedal striding (White & Suwa, 1987).

Looking back into the past, beyond the point when organic material (e.g. animal and plant soft tissue) has deteriorated to unrecognisability, we are left with stones and bones and sometimes DNA. The former must be recovered in the archaeological and palaeontological records (Panger et al., 2002a), while the latter must be recovered from exceptional circumstances of preservation, such as permafrost or museum cupboards (Krings et al., 1997).

Whatever the difficulties, the culture of an array of extinct hominins is there to be inferred (Foley, 2002). At 1.8–1.0 million years ago there are digging tools of bone polished as if used to excavate termite mounds (Backwell & d'Errico, 2001). There are 2.5 million-year-old cutmarks on bones from the use of flaked stone tools used in butchery (de Heinzelin et al., 1999). Later, at 2.0–1.5 million years ago there are ungulate long bones smashed open for their marrow, using stone hammers and anvils comparable to those used by living apes (Bunn, 1981; cf. Blumenschine & Selvaggio, 1988). At about 800 thousand years ago, there are pitted stone anvils that resemble the ones used by living chimpanzees to crack open nuts (Goren-Inbar et al., 2002). In none of these cases can we say who used these tools – apes or humans, or both – but in all these cases we interpolate between what we know of living apes and humans, in seeking to understand our ancestors (e.g. Yamakoshi, 2001). Short of a time machine, what else can we do?

Put another way, whether we like it or not, we must do palaeo-cultural anthropology, and it will have to be informed by both cultural primatology and cultural anthropology.

Archaeologists know this already, in doing ethoarchaeology on living nonhumans as well as ethnoarchaeology on living humans. Sept (1992, 1998) showed that the overnight sleeping platforms (nests) of chimpanzees may shed light on the ranging of hominins over the landscape. Pickering and Wallis (1997) showed that bones chewed by chimpanzees leave distinctive marks that may help us to distinguish between consumption by carnivores versus hominins. Plummer and Stanford (2000) showed that the bone assemblage left behind by chimpanzees after the hunt of a colobus monkey can be analysed in the same way as a hominin kill-site (see also Tappen & Wrangham, 2000).

But this is not enough, given recent findings on behavioural diversity (Whiten *et al.*, 1999, 2001) and an archaeological record (Mercader *et al.*, 2002) for chimpanzees. Now we must do ethno-etho-archaeology: we must seek for clues about hominisation in the cultural diversity, material and otherwise, of living apes.

Writing about these issues in 2003 is a far cry from that personally revelationary day in January, 1975 at Mahale. We have come this far only because again and again, researchers studying apes have had to expand their viewpoints and paradigms. If this chapter has caused you to allow even for the possibility of cultural primatology, then please read on.

2 Definition

'That complex whole which includes knowledge, belief, art, morals, custom, and any other capabilities and habits acquired by man as a member of "society"'

(Tylor, 1871)

'The way we do things'

(McGrew, 2003b)

There is no consensual definition of culture, even in anthropology, which invented the idea. Further, no mainstream definition of culture has been conceived with a view to tackling the main issue of this book: the prospect of humanlike culture existing in nonhuman species. Instead, it is usually assumed in anthropology that culture is uniquely human, as discussed in the last chapter, and as exemplified by Tylor's classic definition quoted above ('acquired by man').

Let us deconstruct Tylor further, as a way of starting to seek a useful definition for the task at hand. His definition seems to have three elements: culture as entity ('complex whole'), culture as content ('knowledge, belief, etc.'), and culture as collective ('member of society').

To take the middle one first, there is a handy list of five specified examples, plus a catch-all closing phrase ('any other capabilities and habits') that suggest that anything and everything human is cultural. This is neat, but is the antithesis of discriminatory: is there anything that humans do, or are, that is *not* cultural, and so might be shared with other species? Apparently not. Further, none of the specified attributes is clearly concrete and observable, and each of them requires definition, too. (Perhaps art is the exception, but it is a truism that art is in the eye of the beholder, not in the object itself.) Knowledge, belief, morals, etc. must be inferred, typically in our fellow humans on the basis of verbal report. Nonlinguistic creatures need not apply, by definition.

This ambiguity makes inferring the presence or absence of a complex whole even more difficult, for presumably such inference depends on being able to ascertain the systemic interaction of the components. Even if we know how beliefs may persist despite knowledge to the contrary, how can we know how these relate to morals? Even with familiar fellow humans, this can be devilishly hard, and even more so with strangers from another culture, or members of another species.

Perhaps one can go further with the third element: membership in society. At first glance, this seems manageable. Who has not seen a huge flock of birds wheel and alight, only to rise again as one? Surely such magnificent choreography must indicate a natural collective. Well, not really. Such intricately co-ordinated group action, however impressive, need be no more than many individuals acting as perceptually reactive automatons, embedded in the geometry of a 'selfish herd' (Hamilton, 1971). Self-preserving, antipredator behaviour is enough to explain the aggregation, without needing to invoke society.

If I have belaboured Tylor, I apologise, but the ubiquity of frustratingly vague, untestable, all-inclusive, and anthropocentric definitions holds. Many introductory textbooks in anthropology resort to catchy epigrams: 'culture is the human ecological niche', 'culture is the human adaptation', 'culture is what humans do'. If we are to see how far the concept of culture can be stretched beyond *Homo sapiens*, we need a different kind of definition.

One tempting solution is to dodge the issue by just giving a 'not quite' label to whatever nonhumans do. Thus, if humans have Culture, then nonhumans might have 'culture,' or protoculture, or preculture, or infraculture, or quasiculture. Unfortunately, rarely are these weasel words defined, either; instead their common gist boils down to the implication of some lesser version of the fully fledged human version. Of course, human culture is unique, but then so is everything else human, just as it is for *any* species. So is chimpanzee culture unique, and orangutan culture, macaque culture, etc. We do not invent such a range of near-miss terms to compare human versus nonhuman digestion or locomotion, so why do so for culture?

CHECKLISTS

Another way to define culture is by checklist, or feature analysis. This means devising an exhaustive list of component features of a complex phenomenon which then amount to a set of criteria. If all items on the list can be ticked off as being present, then it must be culture. Hockett (1960) presented such a scheme for language, with eighteen features, ranging from directional receptions to syntax (Snowdon, 2001). (Actually, he did the same for culture, too, although this is cited less often (Hockett, 1973).)

The first such checklist for culture, applied to a nonhuman species, was inadvertent and implicit. When asked if Köhler's (1927) striking new findings on chimpanzee mentality meant that the apes were cultural creatures, Kroeber (1928) devised a test. He chose dance as an indisputably cultural expression, and then posed what amounted to six conditions: innovation, dissemination, standardisation, durability, tradition, and diffusion. If apes showed all six features, then they could join the culture club.

Table 2.1 shows how chimpanzees do on the Kroeber criteria (see also McGrew, 1992, pp. 76ff). There are multiple examples for each feature, but no single behavioural pattern can be ticked off for all six. In contrast, McGrew (2003a) shows that black rats stripping pine cones meet all six criteria (see Terkel, 1996, for original research). At first glance, this is disappointing, given that some studies of chimpanzees have gone on for decades. The problem lies in the first and last features: innovation and diffusion.

Innovation is not rare in wild primates (Reader & Laland, 2001) but inventions that catch on are. For every novel behavioural pattern that emerges and is passed on to others, scores probably never spread beyond the inventor, and thus fade away. Also, novelty can be recognised only against a backdrop of its known previous absence. Assured recognition is not easy, especially for rare behavioural patterns shown by long-lived creatures.

Diffusion requires emigration and immigration by a culture-bearer, which requires for observational purposes at least two habituated, neighbouring communities of chimpanzees. One community

Table 2.1 *Kroeber's (1928) conditions of culture exemplified in wild chimpanzees.*

Criterion	Characteristic	Example	Source
Innovation	New pattern invented/modified	Pestle pound	Yamakoshi & Sugiyama, 1995
Dissemination	Pattern acquired from another	Social scratch	Nakamura & Uehara, 2004
Standardisation	Form of pattern consistent/stylised	Ant dip	McGrew, 1974
Durability	Pattern performed when other absent	Leaf sponge	Goodall, 1986
Tradition	Pattern transmitted across generations	Nut crack	Matsuzawa, 2003
Diffusion	Pattern spreads across groups	Grooming hand-clasp	Nakamura & Uehara, 2004

has to yield a transferring female, usually a young adult, and another community has to receive her. Few field studies until recently could follow such emigrants and so monitor diffusion, as most field studies have concentrated on one community only.

Termite-fishing, the use of flexible probes of vegetation to extract termites from their earthen mounds for food (Goodall, 1963), is an instructive example. The complex of motor patterns apparently is transmitted, showing *dissemination*. The techniques of harvesting the insect resource and the raw materials involved are consistent and even stylised: Gombe chimpanzees prefer grass stems, while Assirik's prefer twigs. Assirik's termite-fishers fastidiously peel the bark from their probes, while Gombe's never do (McGrew, 1992, p. 168 ff.). Thus, there is *standardisation*. Young chimpanzees learn to fish for termites literally at their mother's knee as she sits on the mound, but by the time they have reached adolescence, when they range alone, they perform the task perfectly well. Thus, they show *durability*. Finally, there are data on up to four generations of termite-fishing from Gombe. For example, from the famous F-lineage, we have data from Flo, Fifi, Fanni, and Fudge. The *tradition* is well documented.

But, like so many other behavioural patterns, termite-fishing was established already when Goodall began the research at Gombe in 1960. Who knows how many years, centuries, or millenia have passed since the technique was invented. Similarly, there are apparently only two communities – Kasakela and Mitumba – in the Gombe population and both fish for termites. Thus, we cannot say whether termite-fishing diffused from one to the other, or arose independently in each.

If this seems disappointing, consider how few cultural patterns shown by humans living in traditional societies could meet all six of Kroeber's criteria. It is easy to think of cultural novelties imposed by global forces (e.g. Coca-Cola or cargo cult), but has any sociocultural anthropologist documented an indigenous example? Definition by checklist is just not good enough, and if doubts remain, what would we do with a chimpanzee group that showed five of the six features? To grant it 83 per cent cultural status seems so simplistic as to be silly.

BEYOND BEHAVIOUR

So far, everything discussed by way of defining culture in terms of comparative analysis has relied on *behaviour*. That is, on overt acts that can be recorded, and so can be counted, sequenced, lumped, or split. Especially with the aid of audio- or videotape, this is the most accessible aspect of the phenomenon. We can transform accents into spectrograms, food-sharing into contingency tables, or mate choice into optimality strategies.

However, a sociocultural anthropologist might say that behaviour is the *least* interesting face of culture. Or that artefacts, as the products of behaviour, are merely concrete manifestations of far more interesting processes. For example, a richer understanding of culture comes from the knowledge that underlies the behaviour.

Culture is really about cognition, not action. It is mind, not matter; conception, not perception.

But even knowledge is not enough. Arguably, the essence of culture is the meaning that is attributed to the knowledge that underlies the behaviour. For example, kinship is shared genes, and its behavioural expression may be in nepotism or incest avoidance. Knowledge of kinship is what makes matriliny easier to implement than patriliny, given internal fertilisation and concealed ovulation in humans. (We can all see who has borne a baby, but who knows who fathered it?) Meaning is what allows kinship to go from genes to roles, whether this be in cross-cousin marriage or mother's brother-to-nephew inheritance.

So, the task is how to define culture in terms of behaviour, knowledge and meaning, so that it can be investigated in other species. More precisely, can culture in the rich sense ever be shown in dumb brutes? Many researchers might throw up their hands at the apparent impossibility of the task, but the solution is simple: we must use the same processes, if not the same methods.

Because we cannot read the minds of our fellow humans, any more than we can read the minds of other species, we rely in both cases on inference. These inferences range from hunch to syllogism, but it all comes down to rational, probabilistic assessment of acts, in terms of antecedents and consequences. We are better at inference when we have large, systematically recorded datasets that cover all the relevant variables than when we have few, opportunistic, impressions gathered *ad libitum*. Whether we engage in discourse or ethology, the more precise and explicit the inference, the better.

Some argue that no definition of culture, human or otherwise, is valid unless it includes symbol use (e.g. Tomasello, 1999). If symbol use means arbitrary assignment of signal to referent (e.g. flag = patriotism), and if access to symbols, or at least to their explanation, is through verbal report, then we are back in a familiar fix. In this view, as articulated on the topic of chimpanzee culture by Washburn and Benedict (1979), language and culture are inseparable. But language as cognition is different from language as communication, in one crucial methodological sense: language in thought is private, while language in speech may be public. We can record the latter, but must infer the former.

So, if in experimental studies in captivity, apes appear able to use nonlinguistic symbols (Savage-Rumbaugh, 1986), and if symbolic thought (but not necessarily symbolic communication) is sufficient to be cultural, then we are allowed to infer symbol-based culture in apes, given apt evidence. In principle, symbolism without language will suffice. (If this seems far-fetched, imagine that we find a remnant population of Neanderthals in the Pyrennees. They utter unintelligible grunts but engrave stick figures on cave walls. We would be likely to grant them cultural status on the basis of their depictions, without waiting to establish their linguistic capabilities.)

ESSENTIALS

Another way to tackle the definition of culture is to look for its essentials. If there is a core set of ideas that all agree on, then this gist could be a necessary foundation for any useful definition. Four candidate essentials are obvious:

(1) Culture is *learned*. That is, it is not instinctive, genetically determined, and hard-wired, although the capacity is innate, epigenetically constrained, and soft-wired.
(2) Culture is learned *socially*. That is, it is not acquired solitarily, whether by simple trial-and-error or complex insight. It is learned from other members of the same species, rather than from the abiotic or biotic environment.
(3) Culture is *normative*. Its expression is not plastic or random, but instead is bounded in space and time, so that the unusual stands out against the usual.
(4) Cultural is *collective*. It is not idiosyncratic, but is instead characteristic of a natural group, whether at the level of community, tribe or nation.

Another school of thought emphasises that a defining characteristic of culture is diversity. In humans, cross-cultural variation is the fabric of the rich tapestry of ethnography. An extreme form of this viewpoint is cultural relativism, which means that all that human beings are and do is a reflection only of their particular societal circumstances. In this view there is no human nature, because there are no human universals (for evidence to the contrary, see Brown, 1991).

This viewpoint is at the core of the first systematic cross-cultural comparison of chimpanzee behaviour. Whiten *et al.* (1999) discounted any behavioural pattern shown by all unit groups that were sampled. That is, behavioural patterns that were present at all seven study sites were discarded from the pool as chimpanzee-universals. Instead, they retained thirty-nine other patterns on the basis of their being well established at one or more sites but known to be absent from one or more other sites. That is, diversity equals culture.

An alternative viewpoint is that culture is shown most convincingly in variations, or even nuances, on common themes. If all human societies employ words for colours, but if there is variation in the scope of these terms, and if that variation is nonrandom and lawful, then we have strong grounds for invoking culture as the explanation for variation. Berlin and Kay (1969) showed that if a human culture had only two colour terms to label the spectrum, then they are designators of black and white. If three terms, then black, white, and red; if four terms, then black, white, red, and green, or yellow, etc.

No one has looked at colour labels in chimpanzees, although given their equivalent colour vision to humans, this would seem to be feasible experimentally. More surprisingly, no one yet has compared systematically the variations in chimpanzees' universal behavioural patterns, e.g. drumming or nest building. Based on tentative analyses (Baldwin *et al.*, 1981, on nests; Arcadi *et al.*, 2004, on drumming), this can, and should, be done.

Recently, Fragaszy and Perry (2003) have argued that tradition is the key to culture, making it both a necessary and sufficient condition for both humans and nonhumans. Depending on how one defines traditions, this could be doubly wrong. If tradition is behavioural continuity over generations, then this is neither necessarily nor sufficiently cultural. Well-worn trails trod by generations of ungulates from valley to valley, or waterhole to waterhole, may be learned individually because they are the optimal solution to a natural set of problems: finding food and water, and energy budgeting. So long as the environment remains constant enough, trial-and-error learning of routes will suffice, without the need for vertical social learning and culture.

Similarly, lots of culture is short-lived or is transmitted by horizontal (peer-to-peer) social learning. For example, Goodall (1968) reported a 'fad' for Gombe's chimpanzees building nests in oil palms that seemed likely to be influenced socially, but it lasted only a few months. Hannah and McGrew (1987) described how the immigration of an adult female chimpanzee with nut-cracking skills caused the rapid transmission of the technique to a previously non-nut-cracking group. Both of these examples occurred within, not across, generations. It seems that there is tradition without culture, and culture without tradition.

Some have argued that any definition of culture must include its cumulativeness, i.e. the idea that new knowledge must build on past knowledge, thus creating history (Tomasello et al., 1993). Tomasello (1999) calls this the 'ratchet effect', implying that cultural evolution is progressive, and says that it is uniquely human. Boyd and Richerson (1996) stressed similarly that truly impressive cultural evolution occurs when behaviour is so complex that it could not have been invented on its own.

This is refuted easily by reference to the continuing elaboration of the two classic behavioural patterns invented by the Japanese macaques of Koshima: sweet-potato washing (Kawamura, 1959) and wheat-sluicing (Kawai, 1965). If one reads only the early reports, then it is no surprise that ratcheting is deemed to be absent. Instead, the techniques have been ratcheting ever since their invention (Kawai et al., 1992; Watanabe, 1994; Hirata et al., 2001a). Some changes were simple: first, the monkeys dropped or threw wheat grains into the water (but there the grains could be pirated by others), so then the monkeys sluiced the wheat in their hands, thus retaining possession of the food. Finally, the monkeys ceased to rely on the wave action at the public shore, and instead began to dig private little flotation pools on the beach.

It may be that human culture accumulates faster than other species', but it is hard to know if it has been doing so for longer. Lacking an archaeological record, we have no idea of the time depth of wild chimpanzee culture (Mercader *et al.*, 2002). It could be decades or millenia. Similarly, while we all would agree that few of us could invent calculus *de novo*, there are many facets of traditional cultures that are not so complicated and could be re-invented by each generation, e.g. food processing.

So, there are many aspects of culture, some of which are likely to appear in most definitions of the phenomenon. But each of these elements is problematic, so it is not surprising that no single definition is accepted by all culturologists, at least as it could be applied to other species. No matter how culture is defined for comparative application, it must be operational. That is, ideas and essences are fine in principle, but in practice, a useful definition must be capable of empirical testing. If the null hypothesis is evolutionary continuity in culture across species – as in other natural processes – then any hypothesis of human uniqueness (to the contrary) must be framed in testable terms. For example, hypothesis: language and culture are inextricable, and only humans have language, so no nonhuman is cultural. Task: falsify either premise, or find culture in a nonlinguistic nonhuman species.

THE WAY WE DO THINGS

By this point, the reader may have become impatient enough to ask: what is the author's definition of culture? After such a reactive critique, perhaps it is time for proaction. My current working definition of culture is the five-word phrase given at the start of the chapter: 'The way we do things' (McGrew, 2003b). This phrase captures four characteristics, which, if present in a creature, qualify it as cultural:

(1) 'Do things' refers to overt acts, that is, behaviour that can be recorded, counted, measured, etc., whether these be dance-steps or words.
(2) 'The way' refers to form in behaviour or artefact that is standardised or even stylised, as opposed to alternatives of other forms, or to lack of form altogether.
(3) 'We' refers to collectivity, i.e. to acts or objects that are socially significant, through meaningful interaction, rather than merely individual behaviour done in parallel with others.
(4) 'The way we do things' refers therefore to the source of a sense of identity, as our way of doing things differentiates us from the way others do things, i.e. 'us' and 'them'.

If we found a previously uncontacted group of people, and if these people showed such action, standardisation, collectivity, and identity, we would hardly deny them culture.

Do chimpanzees qualify as cultural by this definition? The first two aspects are easy, but the latter two are more difficult. Wild chimpanzees do things that look cultural: they show behaviour that is varied, apparently as a result of social learning, across a variety of levels, from clan to subspecies. These behavioural patterns cover a wide variety of functions from subsistence, to self-maintenance, to sociosexual life. We have massive datasets to document this: Whiten *et al.*'s (1999) article in *Nature* was based on a collation of more than 150 years of field study.

Wild chimpanzees also do things that are standardised, and even stylised. Ant-dipping at Gombe is done the two-handed way (McGrew, 1974), at Taï, the one-handed way (Boesch & Boesch, 1990), and at Bossou, both techniques are used (Humle & Matsuzawa, 2002). This difference in technique is reflected in the artefacts employed: two-handed dipping wands are so much longer than the one-handed dipsticks that it is possible to classify a population and to infer its technique with only the tools to go on. Thus, the unhabituated chimpanzees of Mount Assirik can be said to be two-handers on the basis of their long wands (McGrew *et al.*, 2003).

To say that a behavioural pattern is stylised is to imply that its standardisation goes beyond function into another realm. It introduces artificial and arbitrary uniformity. The grooming hand-clasp of Mahale shows this. The original description in K-group was of a symmetrical, elegant mutual gesture, with a palm-to-palm configuration that maintained the fully extended arms overhead (McGrew & Tutin, 1978, see Figure 1.1 of this text). The extreme stylised version is the non-palm-to-palm version in which no clasping occurs at all; instead, the wrists of the two participants merely touch, making minimal contact (see Figure 2.1). Figure 2.2 shows this in close-up view, in M-group. Such performers seem to be going through the superficial motions of the gesture.

Collectivity in chimpanzees requires inference, just as it does in humans. If you walk down the street of a European capital city on a Saturday afternoon, you may meet a noisy mobile group of mostly young men, clad in scarves and hats of the same colour. One will start a chant, and the others will join in. We infer collectivity in these football supporters because they act in unison and pursue a common goal, but more so because they show roles and conventions. (For an insightful evolutionary analysis of this phenomenon, see Morris, 1981.)

A party of male chimpanzees on patrol shows similar characteristics. Their locomotion is stealthy and deliberately paced. As they approach the boundary of their territory, they more often reassure one another with touches, even embraces. Once neighbours are detected, the patrol's manner changes to fear or aggression, depending on relative numbers in their and the neighbours' parties (see Wrangham's, 1999, imbalance of power hypothesis). If a neighbouring male is subdued by three or more resident males, he is likely to be injured severely, even killed, by their co-operative attack. The attackers show division of labour: some hold him down on the ground by arms and

FIGURE 2.1 Non-palm-to-palm grooming hand-clasp at Mahale. Photograph by Michio Nakamura.

legs, while others slash and pummel him. (For a gory case study, see Muller, 2002.)

A cleaner example of role playing in chimpanzees is the alpha male role in M-group at Mahale, especially as expressed in the distribution of meat after the kill of a red colobus monkey. Nishida, *et al.,* (1992) described the impact of the alpha male Ntologi's meat-sharing strategies on the whole unit group. He showed complex, differential sharing of meat to allies, rivals, and neutrals, that, if seen in humans,

FIGURE 2.2 Close-up of wrist-to-wrist hand-clasp grooming of M-group, Mahale. Photograph by Linda Marchant.

would be termed 'Machiavellian'. Ntologi's successor, Nsaba, was more hamfisted, in that he shared meat with only one other male, Kalunde (another ex-alpha), and this dyadic oligarchy left all other adult males out of the process (Marchant, 2002). Thus, alpha male-ship assumed upon achievement of the top rank and lost when another usurps it has a pervasive impact on collective, everyday, life in the unit group.

Chimpanzees differ as much in physical appearance and manner as do humans, as anyone who has worked with the same apes over a long period will say. Thus, we can identify them readily as individuals and give them names to aid our record-keeping. This is identifiability, and it is widespread in vertebrates.

Chimpanzees also give evidence of self-identification, by being able to pick out their own photographic images (Hayes & Hayes, 1954) or to recognise themselves in mirrors (Gallup, 1970). This ability is shared only with humans, other great apes and certain cetaceans, and such self-conception is indicative of identity in the individual sense. By this reckoning, a chimpanzee has a persona.

But what about a social identity? Can an individual conceive of itself as part of a group? Arguably, this is what makes collectivity interesting. The simplest example is xenophobia, the tendency to divide the world into 'us' versus 'them'. The fatal encounters between patrols of males described above are a stark manifestation. But more is required than the simple mentality of 'love familiars and hate strangers'. Membership as a common social identity is acquired socially and can be lost.

Consider the case of Goliath, one-time alpha male of the Kasakela community at Gombe (Goodall *et al.*, 1979). As such, he was familiar to all, but then he joined the secessionist Kahama community that split off to the south. For a while after leaving, he returned occasionally with impunity to the territory of his former Kasakela comrades, but eventually they found him alone and killed him. Somehow, over time, he had changed from 'one of us' to 'one of them', and paid the price.

A less drastic indicator of social identity comes with immigration. While male chimpanzees are philopatric and xenophobic, females are normally the dispersing sex. Males are born and live their lives in one territorial community. At around the time of sexual maturity, females typically leave their natal community and transfer to another community to mate and to breed. Thus, the life history of a female chimpanzee entails an identity change. But how exactly is this transition shown?

Few females have been followed from emigration to immigration, so systematic 'before versus after' data are scarce. However, two females now at Mahale are known to have moved from K-group to M-group, and one of them (Gwekulo), was seen to do the grooming hand-clasp in both groups (Nakamura & Uehara, 2003). In her new group (M), she adopted its flexed elbow style, showing behavioural adaptation to her new colleagues, but she also maintained her old group's (K) palm-to-palm clasp, unlike the others in her new group.

Thus, Gwekulo's maintenance of an identity marker from former times was evident. (Matsuzawa, 1994, has used the similar performance of a foreign type of nut-cracking by an adult female, Yo, at Bossou as a diagnostic of immigration.)

In closing this chapter, I return to the first point made: that there is no consensus in defining culture. To seek after unanimity smacks of a fruitless quest of some holy grail. The key to a heuristic definition of culture is that it must be clear, explicit, and operational. So long as each student of the phenomenon makes it plain what is meant by his or her use of the term, others are free to accept or reject the idea. If culture is an emergent entity, then some inference always will be necessary. The reader may not agree that culture is 'the way we do things', but at least I have tried to make clear my standpoint.

3 Disciplines

Some of the most interesting phenomena defy the boundaries imposed upon them by academicians. Instead, these entities send out tentacles that cross disciplinary divides, or bridge even greater epistemological barriers, such as art versus science. Thus, language engages more than linguists, intelligence more than psychometricians. Culture is one of these phenomena, and its cross-disciplinary (but not necessarily interdisciplinary) nature can be seen as a mixed blessing.

At least four traditional academic disciplines have a significant intellectual stake in the concept of culture (McGrew, 1998). Each of them asks different sorts of questions and uses different sorts of methods to answer them (see Table 3.1). *Anthropology* mostly asks '*what*' questions, addressing issues of phenomenology. *Archaeology* mostly asks '*when*' questions, addressing issues of history. *Psychology* mostly asks '*how*' questions, addressing issues of mechanism. *Zoology* mostly asks '*why*' questions, addressing issues of adaptation. If culture is to be explained, then arguably all four disciplines must make their contributions (McGrew, 2001a, 2003b). The aim of this chapter is to explore what those contributions have been and might be, especially in cultural primatology.

ANTHROPOLOGY

Anthropology invented the culture concept, but no longer owns it. With its inception in the nineteenth century came priority, and so jurisdiction for anthropology (Kuper, 1999). Culture remained the core concept of the discipline for at least a century, and even now, most of the membership of the American Anthropological Association label themselves as cultural anthropologists, outnumbering all other sub-fields combined. Anthropology has been defined epigrammatically as

Table 3.1 *Four disciplines that contribute to understanding culture.*

Discipline	Type of Question	Key Element
Anthropology	What?	Phenomenon
Archaeology	When?	History
Psychology	How?	Mechanism
Zoology	Why?	Adaptation

the science of culture, and most introductory textbooks include some variation on this theme in their characterisation of the discipline.

Two things have happened to change this neat arrangement: recently many cultural anthropologists have questioned or even abandoned the concept, and, roughly over the same timespan, nonanthropologists increasingly have appropriated it (Alvard, 2003). These two opposing tendencies seem to be independent, but they may interact, if anthropologists are keen to jettison the culture concept, yet reluctant to let others make use of it (Ingold, 2001).

Why does anthropology, or at least some proportion of anthropologists, consider the culture concept to be outmoded or even unnecessary? Perhaps the concept is simplistic, or worse, it smacks of Western hegemony over subjugated indigenous peoples, or of intellectual imperialism and accompanying implicit racism and sexism. Phenomenologically, the culture concept may imply an objective reality that is independent of social construction, and thus something that can be studied, rather than only intuited by engaging in discourse. These, and many other issues, go far beyond the bounds of this book (but see Kuper, 1999; Borofsky *et al.*, 2001).

The idea that culture might extend beyond humanity to other species is not new in anthropology, regardless of the frequent claims of its human uniqueness. Lewis Henry Morgan (1868), who founded ethnology in United States, wrote a monograph on the technology of the beaver, as well as studying Iroquois kinship. Ruth Benedict (1935), in her classic, *Patterns of Culture*, asserted a graduated, not abrupt, transition from nonhumans to humans. Marvin Harris (1964) concluded that other species had their cultures, and that the differences between human and nonhuman culture were those of degree and not kind. Similar questions were asked in sociology (Hart & Panzer, 1925). Thus, the current resurgence of interest in nonhuman culture is only the latest in a tradition of such enquiry (McGrew, 1998; Rendell & Whitehead, 2001; Panger et al., 2002b; van Schaik et al., 2003).

It follows that many persons who pursue cultural primatology will follow, intentionally or not, precedents set by cultural anthropology. Just as the latter passed historically through stages of natural history, ethnography, and ethnology, so has cultural primatology. When Goodall (1973) first raised the issue of behavioural diversity across populations of wild chimpanzees, she cited opportunistic, descriptive, data from a few sites. When Whiten et al. (2001) compared systematically nine populations of chimpanzees across Africa – from Senegal to Uganda – they drew upon databases totalling almost 170 years of study.

Explanatory concepts that have proved useful in cultural anthropology are doing so again in cultural primatology. It seems that the distribution of certain patterns of elementary technology can be explained by diffusion, e.g. the origin of nut-cracking in the 'three corners' area of Liberia–Guinea–Ivory Coast, with its eastward spread halted by the zoogeographical barrier of the Sassandra-N'zo river (Boesch et al., 1994). Other patterns seem to have had multiple independent inventions, e.g. termite-fishing seems to have emerged at least three times, in eastern, central, and far western Africa (McGrew et al., 1979, McGrew, 1992).

Other anthropological concepts have the potential to clarify matters in cultural primatology. *Enculturation* is the normal acquisition of knowledge and skills through socialisation. It applies readily to a chimpanzee daughter's learning extractive foraging skills from her mother, for the two are inseparable for years during the daughter's infancy and childhood. But this will not explain her son's acquisition of male skills, such as agonistic display sequences, which the mother never performs. *Acculturation* is the adoption of the knowledge and skills of an alien culture. It is likely to occur when a female chimpanzee emigrates from her natal group and adopts the gestural patterns of the group into which she immigrates. But does this mean that acculturation is sex-specific to chimpanzee females, as males are philopatric and so do not disperse? And what are we to make of ape infants who are foster-reared from birth by human caretakers in isolation from their biological species (Temerlin, 1975)? When these home-reared apes show human behavioural patterns, it is hard to say if this is enculturation (through honorary humanhood) or acculturation (across species, not cultures).

A sign of a discipline's intellectual vigour may be when terms and concepts burgeon too fast for consistency to be maintained. Cultural primatology already has two meanings: the one used here refers to primatologists who specialise in studying cultural, as opposed to physiological, ecological, etc., primatology (McGrew, 1998); the other sense of the phrase refers to the study of interaction, usually traditional in primate habitat countries, of human and nonhuman primates at local level (Fuentes & Wolfe, 2002). This might more usefully be termed 'ethnoprimatology' (Wheatley, 1999). This may range from dietary taboos by hunter-gatherers to sacred status that affords protection, e.g. *Hanuman langurs* being protected by Hindus in India.

Whether cultural anthropologists interested solely in *Homo sapiens* will reach out to cultural primatologists, as colleagues focused on a common phenomenon – culture – remains to be seen. It may be that habits of presumed human uniqueness are too strong to overcome from both sides. It may be that the contrasting methods of enquiry (i.e. verbal report versus observation) are too different, especially with the inherent asymmetry. (We can always observe humans if we choose (e.g. Eibl-Eibesfeldt, 1989), but cannot interview apes.) Cultural primatologists have a great deal to learn from cultural anthropologists, either to make use of their successful ideas or to avoid their costly lessons learned, such as ethnocentrism. Whether or not this will happen depends on goodwill and some risk-taking by both parties.

ARCHAEOLOGY

Archaeology supplies the time depth to culture. Without it we would be stuck in the present, or at least in the oldest living memories of elders. Without palaeoanthropology, we have no sense of origins; without historical archaeology, we have no real sense even of our great-grandparents, much less our earlier predecessors, unless we are lucky enough to live in a literate society with written records. However, in seeking to reconstruct the past, we are bound by the material, no matter how rich is our interpretation of the artifacts (Boivin, 2003).

To understand what this means to primatology, consider termite-fishing, the first type of elementary technology seen by Jane Goodall (1963) less than 45 years ago. The custom was well in place when she discovered it, but it *could* have been invented in 1959, for all we know. That would make its time depth merely decades, but it could be that Gombe's chimpanzees have been fishing for termites for centuries, or millennia, or even for millions of years, right back to the time of the last common ancestor (LCA) or even before. Similar possibilities apply to the termite-fishers of Assirik in Senegal (McGrew *et al.*, 1979) or of Lossi in Congo (Bermejo & Illera, 1999). Without archaeology, we are stranded in the here and now.

Because archaeologists will never see their subjects in action, all of their data on past cultures are inferential and indirect. This is a daunting constraint, because all archaeological progress must be based on inference and can never be validated completely. Archaeology has responded to this challenge with ingenuity and imagination, and their advances have important implications for cultural primatology, *even in the present.*

The reason for this is simple: of the forty or so populations of wild chimpanzees that have been studied, fewer than ten are habituated fully. That is, for only a handful (Bossou, Budongo, Gombe, Kanyawara, Mahale, Ngogo, Taï) are the apes consistently accessible for observation at close range. All the marvellous television documentaries of apes in action come from these few sites. Gaps remain: not a single community of chimpanzees from central Africa, the species' heartland, has been fully habituated. A further handful of populations have been habituated partly: some can be identified individually and sometimes watched. But most of the wild chimpanzees so far contacted by primatologists barely are habituated at all: they flee upon encounter and avoid humans (sometimes with good reason!) even after years of study.

The solution to this problem is the same for primatology and archaeology. In the absence of direct behavioural data, seek information in artefacts and other *products* of behaviour. For primatologists, this means not only tools, but sleeping platforms (shelter), food remnants (processing techniques), drumming sites (communication), trails (ranging), and faeces (diet, parasites, reproductive state). All of this circumstantial evidence can yield knowledge, if the inferences are sound (Marchant, 2003).

For example, Baldwin (1979) hypothesised that the chimpanzees of Assirik used only anvils of stone or wood to crack open the hard-shelled fruits of the baobab tree. In contrast, Bermejo *et al.* (1989) hypothesised that the same chimpanzees used hammers *and* anvils of stone to process these food items. How to choose between these two alternatives, if behavioural data were lacking? Marchant and McGrew (2004) used quasiarchaeological techniques based on the relative position and characteristics of fruits and stones to show that hammers are not necessary to account for the remnants of processing left behind by the apes. At the same time, Marchant and McGrew (2004) were able to validate the 'anvil only' hypothesis by analyses of archival behavioural data from Assirik.

Interaction between archaeology and primatology is not just a one-way street in which the former informs the latter. As introduced in Chapter 1, palaeoanthropologists have begun to make use of primatological findings to interpret their artefacts and residues. Plummer and Stanford (2000) compared the fragmented bone assemblages left behind after a meat-eating episode by wild chimpanzees and early hominids. Tappen and Wrangham (2000) scrutinised bone fragments of colobus monkeys consumed by Kibale chimpanzees, having passed through the ape's digestive tract. Backwell and d'Errico (2001) used knowledge gained from chimpanzee extractive foraging techniques to get termites as a model to reinterpret the function of early hominid digging tools. Goren-Inbar *et al.* (2002) inferred that 800 000-year-old pitted stone anvils from Israel were the nut-cracking sites of early hominids, based on the comparable lithic technology of wild chimpanzees.

The ultimate in productive collaboration between archaeology and primatology in the pursuit of culture is when practitioners from both fields work side by side. For example, Wynn (archaeologist) and McGrew (primatologist) (1989) reasoned in principle that the cognitive abilities underlying chimpanzee behaviour seen in nature would be enough for the ape to make the simplest flaked-stone technology (Oldowan). Schick and Toth (archaeologists) joined Savage-Rumbaugh, Sevcik and Rumbaugh (primatologists) (Toth *et al.*, 1993) to test this idea in practice; Kanzi, a captive bonobo, made stone flakes with a functional cutting edge. Archaeologist Sept, (1992, 1998) joined primatologists Steklis and Gerald in the field in Congo (Zaïre) to study the spatial distribution over the landscape of the simple overnight shelters of chimpanzees. Joulian, an archaeologist, (1994, 1996) turned primatologist with regard to hammer and anvil use by chimpanzees to crack nuts in Ivory Coast.

All of the above are examples of ethoarchaeology, i.e. the study of the behaviour or behavioural products of living nonhuman species as proxies for extinct ones. On phylogenetic and geographical grounds, the chosen models are usually the two species of African great ape – bonobo and chimpanzee – that are most closely related to humans. But, of course, living apes are not extinct apes, much less extinct hominids in the evolutionary lineage that led to humans, after the split from the LCA. The debate on such modelling is ongoing, e.g. Moore (1996); Panger *et al.* (2002a). What is needed is not just actualistic archaeology on apes, but also actual archaeology done on the artefacts and remnants left behind by ancestral apes. Tutin and Oslisly (1995) recognised this need some time ago.

Happily, Mercader *et al.* (2002) have now made a start: using standard archaeological techniques such as radiocarbon dating, they have excavated chimpanzee nut-cracking sites in the Taï Forest of Ivory Coast. Contrary to expectations, they found many flakes of stone that resembled components of Oldowan technology. This strengthens an evolutionary scenario in which anvil use by apes to smash fruits can lead, step by step, to flaked stone technology (Marchant & McGrew, 2004).

Mercader *et al.*'s (2002) ground-breaking study is important for several reasons. First, it shows that archaeology can be done on non-human artifacts; thus, it takes their elementary technology out of the present, and back into the past. Second, it shows that on-site, prospective collaboration between archaeology and primatology can be fruitful. This should help future projects overcome disciplinary barriers in seeking research funds. Most importantly, it recognises that there are at least two archaeological records to be deciphered: ape and human. The challenge now is to find species-specific signatures that distinguish these records. Primatologists knew of such confusion, when it became clear that in the African forests today, both apes and humans crack nuts with similar techniques (Kortlandt & Holzhaus, 1987). Now, archaeologists must face up to the challenge.

Finally, cultural primatology has implications for archaeology in another way: ethnoarchaeology now means looking at the relevant cultural activities of living humans, typically hunters and gatherers as a way to help interpret the human archaeological record. Such ethnoarchaeology can also be done on apes: Pickering and Wallis (1997) got captive chimpanzees to chew on ungulate bones slathered in peanut butter! Their teeth marks may be diagnostic of those of ancestral hominoids, versus carnivores.

An influential synthesis was Isaac's (1978) formulation of the evolution of food sharing, using ethnographic knowledge of southern African, arid-country foragers. But he also cited ethoarchaeological findings from chimpanzees. Twenty-five years later, it is harder to keep the two approaches separate. We know now that at least some of the elementary technology of humans and chimpanzees is very similar: for food-getting, Tasmanian aborigines and Tanzanian chimpanzees have toolkits that differ little (McGrew, 1987, 1992). Further, now that we have verified systematically behavioural diversity in chimpanzees (Whiten *et al.*, 1999, 2001), the distinction between etho- and ethno- as prefixes for archaeology becomes hazy. If apes have varied material culture, then we can no longer simply use the species as the unit of modelling ancestral humans. We must now be careful in choosing which population of chimpanzees to act as proxy in our 'evolutionarios' (evolutionary scenarios). Reliance on Goodall's famous apes at Gombe would be inappropriate for modelling the evolution of lithic culture, as they use few stone tools and do no nut-cracking. We have as yet just scratched the surface of useful co-operation between palaeoanthropology and primatology.

PSYCHOLOGY

Apart from the somewhat marginal subdiscipline of cross-cultural psychology, psychologists have shown little interest in culture until recently. Instead, it is easy to gain the impression that all humanity lies in the responses of first-year undergraduates at North American universities, and who seem to be the main source of data for scientific psychology. However, interest in the possibility of culture in nonhumans, and in the ways that psychology might contribute to the debate, has emerged experimentally in comparative and developmental psychology (Galef, 1992; Tomasello *et al.*, 1993; Premack & Premack, 1994; Tomasello, 1999; Whiten, 2000; Laland & Hoppitt, 2003).

A closer glance shows that these psychobiologists' interest is usually in social learning, and more precisely, in the mechanisms of information transmission that go on in social learning, so that the state of knowledge (as reflected in the behaviour) of the learner is changed. To the extent that psychologists go beyond social learning to issues related to culture, it is often to distinguish between traditions in nonhuman animals versus culture in human animals. (It is curious to ponder why 'tradition' is an acceptable anthropomorphism, while 'culture' is not, in psychological quarters.)

I have addressed in Chapter 2 the pitfalls of hitching the cultural wagon to tradition, but what about the core issue of social learning? A clear definition would be helpful, and would seem to be straightforward, but alas, is not so. A longstanding and prominent critic of these issues recently has defined social learning as: '. . . a general term referring to several behavioral processes that allow social interactions to bias what individuals learn' (Galef, 2003, p. 74). Yet this immediately is suspect, for it rules out a well-known type of social learning called *passive observational learning*, which is common in both animals and humans (cf. de Waal, 2001). The point about such surreptitious information gathering, even eavesdropping (Whitfield, 2002), is that there is *no* interaction between model and learner. In fact, discreet learners may have good reasons to *avoid* interacting with the knowledgeable individuals whom they seek to parasitise. Imagine going to a party and realising that there is a new dance step in fashion. Many people seek to update their moves while trying to hide their ignorance, in order to avoid embarrassment.

For the purposes of this book, social learning will be defined as *social influences on learning*, in which the variable of sociality *derives from a member of the same species (or surrogate) or its products*. Thus, a chimpanzee mother whose nut-cracking affects her offspring's proficiency at the task is an example. But so is a human foster mother whose adopted ape infant comes to walk bipedally. Even the well-worn depression in a stone anvil, created by generations of chimpanzees before, that enables the naïve nut-cracker to be more productive, is an example. What is *not* social learning is individual knowledge or skill acquisition that happens to be done in a social setting. So, coincidental individual learning in parallel (e.g. learning to seek shade on a sweltering day) need not entail any social influence. Similarly, contagious yawning in a stuffy seminar room shows social influence, but none of the yawners is learning to yawn, so it is not social learning.

However, as social learning is a necessary condition for culture, psychology's contribution to understanding the variety of mechanisms of social learning is vital. Whiten and Ham (1992) outlined seven types, from simple exposure ('By being with A, B is exposed to similar learning environment') to goal emulation ('B learns from A the goal to pursue'). Furthermore, it is only by experimentation in captivity (versus even the most systematic and careful observation in nature) that these competing mechanisms (Is it stimulus or local enhancement?) will be elucidated. Field workers who collect data in the rich, uncontrolled setting of the wild may have ecological validity on their side, but they can say nothing conclusive about the causal relations between independent variables. Even if field workers can show strong correlations between variables, rarely can they hold all other factors constant or assign subjects randomly to a particular treatment.

For example, degree of genetic kinship may correlate with degree of similarity in behaviour but, almost always in nature, kinship is confounded with familiarity, so either variable could explain the result. Suppose that at Gombe, Fanni shows similar maternal techniques to her mother Fifi or to her grandmother Flo. This may be because these habits were acquired by social learning from generation to generation, or it may be that there is matrilineal inheritance of personality traits that lead to good mothering. We field workers will never know. In the laboratory, we could, at least in principle, do a counter-balanced cross-fostering experiment to tease apart the parameters of nature and nurture.

IMITATION

Psychologists considering the possibility of culture in nonhuman species seem to stumble on two cognitive factors that are often considered unique to living *H. sapiens*: true imitation and teaching. In true imitation, the naïve individual learns directly about a novel behavioural pattern from another individual who performs it in the learner's presence. This requires that the imitator stores a visual representation of the pattern in its brain and then matches its motor output to that stored representation (Galef, 2003). Such true imitation can be contrasted with all of the other less cognitively demanding processes, e.g. emulation. In theory, this sounds good, but in practice the covert representation must be inferred from the overt actions of the subject. The question of which nonhuman species are capable of imitation is controversial. For an optimistic account, see Whiten *et al.* (2003a); for a pessimistic account, see Visalberghi and Fragaszy (2002).

The neatest way to study imitation is to use the 'two action' experimental design. Here, subjects are presented with a problem to solve in the form of a device with two ways to be opened. The alternative methods of solution differ only in the way that the same features of the device are manipulated. Each naïve subject sees a demonstrator perform only one of the methods; later when tested, if the subject uses only the method seen, then imitation can be inferred. (If the task were being solved by emulation, then either method of solution would be shown by the learner, as emulation deals only with ends, not means.) Whiten and his students (2003a, b) have shown that even young chimpanzees use such imitation successfully in opening 'artificial fruits'.

But even if imitation were proven to be unique to *H. sapiens*, that would say nothing about the question of culture being uniquely human. No anthropologist has ever claimed that human cultural transmission occurred only by imitation. In fact, just the opposite is the case: ethnologists stress the flexibility shown by humans in both enculturation and acculturation. Ethnology has never had an experimental epistemology, and so could never be based on proven imitation. Instead, when cultural anthropologists watch how real people in the real world of traditional societies learn their skills in daily life, it is often impossible to say whether they do so by exposure, enhancement, emulation, imitation, etc. For example, Hewlett and Cavalli-Sforza (1986) documented the learning of the fifty most important skills in the daily lives of Aka pygmies in Congo. The scientists could say little about the specific mechanisms involved, except that social learning was involved.

TEACHING

The case for teaching as an essential basis for cultural transmission is different. If teaching is defined as the *intentionally directed transmission of information from a more informed individual to less informed one*, then it is clearly more complex and problematic. (Caro and Hauser (1992), devised a more lengthy but inclusive definition for the phenomenon that is admirably suited for empirical testing.) It is more complex because the tutor and pupil must interact, rather than one merely reacting to the other. Thus, we must attend to a dynamic feedback loop of the tutor's tailored performance and the tutee's contingent improvement in skill, or enhancement of knowledge. Having teaching as a necessary condition for culture is problematic for two reasons, each equally daunting. One implication of teaching is that it requires theory of mind, i.e. the ability to impute ignorance or knowledge to the mind of the learner: ('I know what you know, or don't know'). This distinguishes teaching from training, with the latter requiring only the delivery of punishment or reward until the learner's behaviour is modified satisfactorily (McGrew, 2003c). In human child-rearing, it seems that we toilet-train, not toilet-teach.

Equally problematic is that teaching, unlike other forms of social learning, clearly is sociobiological (McGrew, 2003b, c). That is, both teaching and being taught incur costs as well as benefits. In social learning, the demonstrator merely goes about her business, and knowledge transmission is the responsibility of the attentive learner. In teaching, the demonstrator must adjust her actions to make them pedagogical; this may entail extra time, effort or risk. Similarly, the social learner is often a reactive 'shopper', able to choose among disinterested models, who may not even know that they are being scrutinised. In contrast, the pupil is engaged with an interactant who may show deception, coercion, etc., rather than the presumed generous sharing of knowledge. Thus, applying Trivers' (1985) matrix of the four possible outcomes to all social interactions, teaching may be *selfish* as well as *altruistic*, *spiteful* as well as *co-operative*.

Needless to say, elevating teaching to a key position in cultural transmission raises thorny issues that are hard enough to disentangle in our fellow humans, much less in other species. Luckily, as with imitation, we are reprieved on two counts. First, there is no conclusive evidence of teaching in nonhumans, though there are suggestive anecdotes (Boesch, 1991c; Guinet, 1991). This should not be surprising, as teaching can be considered more of a curse than a blessing. Because of its costs, both direct (e.g. time consumed) and indirect (e.g. misleading propaganda), it should be invoked only as a last resort, when simpler types of social learning fail (McGrew, 2003a). So, for nonhuman teaching, there is nothing yet to be explained.

Second, as with imitation, teaching is not essential to culture. However important teaching may be in industrialised human societies with schools, it is inconspicuous in traditional ones. Lots of human culture just gets passed on passively and unconsciously. In evolutionary terms, teaching may have been a by-product of literacy. Thus, teaching may be a sufficient condition for imputing culture, but is far from a necessary one.

So, psychologists interested in the possibility of nonhuman culture seem to be caught in a paradox: well-designed experiments in the laboratory produce clear results, but the results may be invalid because of the artificial conditions of captivity. (Why should chimpanzees be motivated to imitate the acts of strange human models, when such a situation is contrived totally?) Results from observations in nature are valid ecologically, but they are confounded hopelessly by entangled variables. (Some populations of wild chimpanzees may ignore nuts as potential food, but is this because they are ignorant, or do they prefer other foods?) Is there a way out of this dilemma?

Actually, there are two – both compromises. One is to do observations in rich and stimulating captive environments. Here, the subjects behave spontaneously, even naturalistically, and at least some variables can be controlled, e.g. life history, and access to resources. Thus, de Waal and Seres (1997) followed the unexpected appearance and spread of the grooming hand-clasp in a captive group of chimpanzees. The observers were able to say with certainty which individuals showed it first, because they had constant access to the apes. This is testimony to the continuing power of the methods of ethology (Tinbergen, 1963).

The other solution is to do quasi-experiments in nature. Although variables cannot be controlled (by definition), some can be manipulated. For example, at a nut-cracking site, one can supply equal numbers of hammer stones in all weight classes, and so see if the chimpanzees show preferences for tools by weight. The foremost practitioner of this approach has been Matsuzawa (2003), with his 'outdoor laboratory' at Bossou, Guinea. There he tests hypotheses about nut-cracking, leaf-sponging and ant-dipping in wild chimpanzees. This is yet another example of experimental field primatology, as pioneered by Kummer (1971).

Psychology's contribution to the debate about culture in other species is both essential and exasperating. It is essential because none of the other disciplines involved will ever be able to tell us about cultural transmission at the level of cognitive processes. Given our big brains and abstract intelligence, we humans have good reason to believe that those mental processes are key to understanding what makes human culture different from that of other forms. But, at the same time, what we know of human cultural diversity comes from cultural anthropology, which has paid almost no attention to such processes. If cultural anthropologists had waited to do ethnography until underlying mechanisms of transmission were clarified, then we would have no ethnographic record, and most of the traditional cultural practices have since disappeared. Thankfully, no cultural anthropologist ever delayed study of kula ring or potlatch or fraternal polyandry because of not knowing if the custom was passed by response generalization or emulation. Knowing the *how* of culture is important, but not knowing it does not stop us from seeking the what, when, and why.

ZOOLOGY

Evolutionary biologists are concerned with the *why* of all creatures, living and dead. Accordingly, they must be interested in culture in any species that shows this adaptation. Unlike social scientists who tend to set humanity apart, natural scientists start with a null hypothesis of phylogenetic continuity. If a prerequisite for culture is capacity to learn, then many organisms are eligible for culture-bearing status, including plants and one-celled animals. But no one has yet found *social* learning in these creatures, so they are ignored from here on (cf. Bonner, 1980). Chapter 4 presents a survey of cultural candidacy across invertebrate and vertebrate taxa.

If the capacity for culture has evolved, it must have been subject to natural selection, or it must be an inadvertent by-product (exaptation) of such selection acting on other traits. (Culture seems too complex to have emerged through the other forces of evolution: mutation, genetic drift, or gene flow.) If culture is a product of natural selection, it could have appeared once in a long ago common ancestor to all living culture-bearers, or it could have evolved several times independently in different lines. If culture is an exaptation (of language, or intelligence, or sociality, for example), then it might be neutral or even maladaptive.

But the capacity for culture is not like most morphological, or physiological, or even behavioural traits. It is at the same time both Darwinian and Lamarckian in its character. Unlike genetic inheritance, cultural evolution can build on phenotypic traits acquired during an individual's lifetime. Furthermore, cultural information can be gained from sources other than one's biological parents and passed on to other than one's biological offspring. Thus, my mother's sister's husband (no genetic kin to me) taught me to fish, and I have taught my students (i.e. academic offspring) how to chase after wild chimpanzees.

There is a huge body of published literature on such *dual-inheritance* theory (see, for example, Boyd & Richerson, 1985). This slim volume is not the place to review it, but something must be said, at least in passing, about the theory of memes. As cultural evo-lution is to organic evolution, then memes are to genes, as the units of information transmitted from individual to individual.

The concept of the meme was invented by Dawkins (1976) and popularised by Blackmore (1999). Like a gene, a meme may affect its carrier's reproductive success, and like genes, memes as replicators have interests that may not coincide with the best interests of their bearers. This makes group selection the norm in memetic evolution, but the exception in genetic evolution. Unlike genes, memes may pass freely among interacting individuals in quick time; these are the fads of popular culture. More problematically, the exact make-up and efficacy of memes is not yet clear: if kissing on alternating cheeks three times is the norm upon meeting another person of the opposite sex, what exactly is the meme: greeting, kiss, number, or heterosexual selectivity? If extemporaneous cannibalism allows an individual to survive an Andean plane crash but gets a person sent to prison for practising it at any other time, what is the efficacy of the cannibalism meme?

What zoology, or, more precisely, evolutionary behavioural ecol-ogy, brings to the table with regard to the question of nonhuman cul-ture is a number of strong, well-tested analytical concepts. First, the comparative approach that allows for the distinction between homol-ogy (similarity by descent) versus analogy (similarity by convergence). This allows us to make sense of differences between closely related species (e.g. chimpanzee versus bonobo, which are sister species as close as horse and zebra), as well as similarities between ape (chimpanzee) and monkey (capuchin), which had a common ancestor only tens of millions of years ago.

Second, optimality theory allows us to cast all traits in terms of the trade-off between costs and benefits, the net result of which impacts reproductive success, or genetic fitness. This applies equally well to gene–meme interaction, e.g. in the evolution of the prolonga-tion of lactase secretion in human pastoralist groups. These societies subsist on the products of their herds, of which milk is the key. Natural

selection has favoured individuals whose lactase production continues after weaning, and this subsistence combination has evolved
several times independently, e.g. in Finland and West Africa (Durham,
1991).

Third, being grounded in evolutionary theory, a biological
approach to nonhuman culture is multifaceted. More to the point,
biological modelling of cultural evolution, and especially the origins
of human culture, is more inclusive than other views. Anthropologists often seem wed to single-species, referential models for ancestral
hominids, whether baboon, bonobo, or chimpanzee (McGrew, 1981).
Most psychologists continue to study primate social learning in dis-
torted social settings, e.g. chimpanzees in constant groups with only
one adult male instead of multimale and multifemale communities
that fission or fuse on a daily basis. Biologists, on the other hand, con-
struct models from basic principles of evolutionary theory, whether
that be mate choice based on sexual selection, or elementary technology based on optimal foraging.

Finally, there is a sociobiological slant to culture. To be useful,
culture must enhance survival and fitness. If knowledge is power, then
we must explain why knowledge is shared at all. Evolutionary theory
offers us reasons for sharing – kinship and reciprocity – and these may
counteract or overrule the obvious problem of innovators (producers)
being parasitised by consumers (scroungers). It may be that teaching is
the adaptive answer to the free-rider problem, i.e. group members who
take the benefits of social living but refuse to pay the costs. Instead
of broadcasting knowledge willy-nilly, natural selection may reward
those who sequester it and then distribute it selectively to family and
friends. This may explain why teaching emerged, for otherwise it is
costly in terms of time, energy, and risks.

It must be obvious by now that a productive cultural primatology requires input from all of the four disciplines covered in this chapter. Merely to follow slavishly the precedents of cultural anthropology
would be to ignore lessons learned and repeat historical mistakes that
have culminated in that discipline rejecting its roots. To follow only

archaeology would be immensely helpful for tackling material culture, but would offer little for direct behavioural or cognitive investigation. To follow only psychology would be to maximise reliability and epistemological elegance with its ensuing intellectual satisfaction, but at the expense of ecological validity. Finally, to follow only biology would make use of the analytic power of reductionism and the most inclusive arena of all (social life), but how could it tackle the emergent properties of such a complex phenomenon as culture?

The next three chapters present the challenges on offer from a wide variety of organisms, all of which stand knocking at the gates of culture.

4   Creatures other than primates

It seems that there is a significant inverse correlation between number of legs and proficiency at social learning. The total biped, *Homo sapiens*, ranks first, followed by the somewhat biped (songbirds), quadruped (nonhuman primates, rodents), hexapod (ants, Reznikova, 2001), octoped, (octopus, Fiorito & Scotto, 1992), and the non-social-learning decapod (e.g. crayfish) bringing up the rear. The trend will be even stronger if those elusive near-nullipeds (cetaceans) turn out to be the most profound social learners of all. Whales do change their songs annually on a global scale (Noad *et al.*, 2000), something that humans cannot match, even with telecommunications.

Is the relation between legs and learning nonsense? Probably, but no more so than some other claims made about culture in animal species other than those of the Order Primates. Definition again is crucial: depending on the definition of culture, only living humans (Premack & Premack, 1994) or even slime-moulds (Bonner, 1980) can be considered as culture-bearers. The former viewpoint is speciest, and exclusive; the latter is exuberantly inclusive, although the altruistic amoeba may feel aggrieved at being omitted (Strassman *et al.*, 2000).

The conceptual confusion may come from the variety of long-standing views on the possibility of animal culture that go back at least to Morgan (1868) writing about the beaver. This hard-working denuder of landscapes and erector of dams and cunning builder of lodges would seem to be the perfect example of a niche constructor (Laland *et al.*, 2000), yet is but a big rodent. Since then, interest in the topic has come from anthropology (Kroeber, 1928), sociology (Hart & Panzer, 1925), psychology (Menzel, 1973b), biology (Bonner, 1980), and philosophy (Dennett, 1996). More recently, cultured animals have captured popular markets, a sure sign of coming of age (de Waal, 2001; Linden, 2002).

So, how to proceed? Laland and Hoppitt (2003) propose an elegant experimental paradigm which, if applied across all taxa, allows for the cultural to be separated from the acultural. In the first experiment, individuals from different populations are swapped across them. If the incomers' behaviour changes to match that of the new hosts', then genetic differences cannot account for the resulting behavioural convergence. But environmental factors could explain it, so a second experiment is needed. In it, two populations are totally interchanged in space, each now occupying the other's former home. If each remains true to its established behavioural patterns, despite being in a different habitat, then social, not environmentally influenced learning is likely. However conclusive in principle, the procedures have been done in practice only with birds, fish, and rodents, and then only in captivity. They will probably not be done with nonhuman primates, because of ethical reservations, and probably cannot be done with pelagic cetaceans. On this basis, Laland and Hoppitt (2003) conclude that there is better evidence for culture in fish than in primates!

On these grounds, what is the evidence for culture in *H. sapiens*? There is certainly no such evidence from sociocultural anthropology, which has never been an experimental discipline. There have been inadvertent 'natural experiments' (e.g. the English troublemakers transported to Botany Bay in Australia), but the converse was never done, and the incomers never were absorbed by local aborigines to adopt their customs. The same applies to Amish in North America, Lebanese in west Africa, Indians in Trinidad, Chinese in Indonesia, etc. And there is no equivalent to the second experiment's wholesale exchange at any time in human history. So, by Laland and Hoppitt's (2003) standards, it looks like the evidence for culture in humans is weak.

Thus, there is a choice: use rigorous experimental methods of comparison but, in doing so, leave out all large-brained mammals, including humans, or use less rigid observational criteria that allow for all major taxonomic groups to be included. The latter is regrettably preferable, until or unless the former conditions can be met. But what less rigid criteria and why? One easy but unsatisfactory answer is that anthropology has never imposed strict conditions for culture. Anthropologists cannot even agree on a single definition for culture, either conceptually or operationally. Therefore, following that precedent, students of cultural primatology need not bother about seeking consensus but, instead, scholars may set their own terms, so long as these are explicit, comprehensive, and clear. This variability is the status quo for field studies of all nonhuman species that are presented as candidates for culture.

A better answer to the question of how to proceed is that different problems call for different definitions. Some researchers seem to have an agenda, implicit if not explicit, to keep membership in the Culture Club restricted to behaviourally modern humans. They compose their definitions accordingly, and equate culture with human culture. Other researchers seem determined to crash the party, speaking as advocates on behalf of their preferred organisms and in opposition to 'primate supremacy' (Bshary et al., 2002; van Bergen et al., 2003). These cultural egalitarians define culture with a low bar, easily leapt. For them, every species in theory can have its own culture by meeting minimal standards. A third set of researchers worries less about definitions and more about labels. They avoid the 'C' word and substitute near-but-not-quite synonyms. Thus, humans have cultures, but nonhumans have traditions. (This designation is entirely arbitrary, of course, as it could just as easily have been the other way round.)

Another reason for labile definitions is methodological: those who see culture as behaviour or its material products (because that is all there is out there to count and to measure) need a different definition than those who see culture as everything but behaviour. These latter-day mentalists see culture as knowledge or meaning, symbolically encoded in the mind, or even more so, in collective consciousness. Because inferred phenomena cannot be observed, they are not bounded by natural law, and so need not be operational.

But it is not so easy to characterise definitions of culture in this way. Here is an erudite, current definition, singled out for scrutiny: 'Cultures are those group-typical behaviour patterns shared by members of a community that rely on socially learned and transmitted information' (Laland & Hoppitt, 2003, p. 151). It starts out sharp but turns fuzzy at the end. At the outset, key characteristics are made clear: behavioural patterns that are: (1) group-typical, (2) shared, (3) socially learned. But what exactly is the socially transmitted *information* upon which the whole enterprise relies? How are we to know it directly?

Having set the stage, the aim for the rest of this chapter is to present examples from nonprimate species that have been nominated as candidates for culture, or at least as social learners. Since social learning is sometimes said to be equivalent to culture, then either will suffice. Then, I will scrutinise the claims in terms of various definitional criteria.

Many invertebrate phyla are credited with the ability to learn, often in simple paradigms (e.g. an individual making a two-way choice in a T-maze), but few are credited with *social* learning. Everyone's humble favourite is the Mediterranean octopus, which has lensed eyes but no real brain (Linden, 2002). An octopus that sees another one in a neighbouring aquarium having a bad experience learns from this to respond appropriately (Fiorito & Scotto, 1992). More impressive are the profoundly more social ants (e.g. *Formica*, *Serviformica*), that work in teams for detecting and retrieving food in foraging. They learn to co-ordinate acts and to find their way home (with appropriate experimental controls for pheromone trails) using social learning (Reznikova, 2001). Whether or not this is more than the well-known one-to-many learning of the location of food sources, as transmitted by the famous 'waggle dance' of honey bees, awaits further investigation.

FISH

Of the poikilotherms, only fish have been put forward as culture bearers (Bshary *et al.*, 2002; Brown & Laland, 2003; Laland & Hoppitt, 2003; van Bergen *et al.*, 2003). (There seems to be nothing published about social learning or culture in amphibians or reptiles.) Brown and Laland (2003) make the strong assertion that social learning is common in fish, and cite five areas of behavioural evidence: (1) anti-predation, (2) migration and orientation, (3) foraging, (4) mate choice, and (5) eavesdropping. Their definition includes the social acquisition not just of novel behaviour, but also of information, so in the latter case, it need not be learning, but may be only perception. For example, eavesdropping is defined as exploiting signals in a communication network, as measured by changes in the observer's behaviour based on which other conspecifics are present.

Van Bergen *et al.* (2003) make much of the primatocentric bias found in students of social learning in animals. They present fish as being under-appreciated and even discriminated against in a world that over-estimates the cognitive abilities of primates (see also Bshary *et al.*, 2002). They ask only that all organisms tested for social learning be subjected to the same criteria, on a level playing field. To illustrate these issues, they cite much evidence from laboratory studies of the guppy. These fish demonstrate perseverance in routes learned from others, even when those routes are suboptimal. They discuss guppy innovation, which is defined as the first subject to swim through a maze. Individual consistency in such innovation is interpreted as 'personality'. The tendency for others to adopt the behavioural patterns of the majority is termed 'conformity'.

On this basis, van Bergen *et al.* (2003) challenge primates as being any better at social learning. They target Imo, the Japanese monkey who invented sweet-potato washing and wheat sluicing (see below). To belittle her, they claim (with little evidence) that 'food washing is a stable feature of macaque behaviour', which will come as a surprise to primatologists. (That a few cases of natural food-washing occur in macaques is not disputed (e.g. Nakamichi *et al.* 1998), but hundreds of other studies have failed to find it (e.g. in free-ranging rhesus macaques).) Van Bergen *et al.* (2003) repeat an old speculation (not a finding) that sweet-potato washing may have been an artefact of human provisioning (Green, 1975). They then assert that Bertha, the maze-solving guppy, was more likely to be a genius than Imo the food-processing monkey. Having made a compelling case for *social learning* in guppies, they then assert that there is stronger evidence for *culture* in fish than in primates.

Evidence for culture in fish in nature comes from coral reefs, as stable habitats with long-lived residents. French grunts use traditional daily foraging routes that depend on social learning (Helfman & Schultz, 1984). Bluehead wrasse have mating sites than remain constant over generations, and experimental replacement does not alter this site-attachment (Warner, 1988).

BIRDS

Of the vertebrates, birds take precedence as candidates for social learning and culture, both historically and in total volume of research. Almost all of the published literature on the subject until recently was on song-learning in passerines. Marler and Tamura's (1964) early work on dialect variation in white-crowned sparrows set the stage. Mundinger (1980) seems to have been the first to set this in a general framework of cultural evolution. There is evidence from captive work on cowbirds that song composition is shaped by the reinforcement of social partners (West, *et al.*, 2003). Also, some analyses of song-learning have made fruitful use of memetics (Burrell, 1998). This is not the place for a review of avian social learning (see, for example, Slater, 1986; Freeberg, 2000), but it is clear that the most sophisticated experimental and field research on culture in nonhuman species has been done on vocal learning in songbirds.

But there is much more to social learning than songs in birds (Marler, 1996). Here are a few examples illustrating key points. Scrub jays show social cognition that is both retrospective and prospective. They use knowledge gained by being past pilferers of others' stored food to modify their present own cacheing strategies. That is, they adjust their behaviour now in light of experience to avoid having their food taken by others (Emery & Clayton, 2001).

On the other hand, perhaps the most famous example of all of elementary technology in birds turns out *not* to depend on social learning. The woodpecker finch of the Galapagos Islands uses twigs or spines to winkle out arthropods from tree holes. This elementary technology is learned individually (Tebbich *et al.*, 2001), based on clearly demonstrated ecological variables (Tebbich *et al.*, 2002).

In a previous book (McGrew, 1992, pp. 212–14), I entered a cautious note about primatocentrism, in recognising that the material culture of bower birds shows rich diversity. Males of some eighteen species make and use structures (bowers) in courtship and sexual display (Humphries & Ruxton, 1999; Madden, 2003). These edifices, which have no other function (e.g. are not nests), range from parallel walls (avenue builders) to spires (maypole builders). Why such elaborate structures have evolved is unclear, but there appears to be an innate component, as evidenced by a correlation between mitochondrial DNA sequences and type of bower (Kusmierski *et al.*, 1997). With regard to putative culture, there is variation in bowers across populations of the same species and across individuals within a population (Diamond, 1986, 1988). Whether this diversity across sites reflects group-level preferences among a range of raw materials (e.g. butterfly wings versus beetle carapaces), or differential availability of these raw materials, remains to be seen.

The most complex evidence of technology in birds comes from the Caledonian crow of the Pacific Island of New Caledonia (Hunt, 1996). It has multiple-tool-use traditions, and the evidence for these is based not on behaviour, or even on artefacts, but on the 'counterparts' where the tools are detached from leaves of the *Pandanus* palm. These leaves tell a history of progressive change in the probing tools made by the birds to winkle out invertebrate prey (Hunt & Gray, 2003). By comparing twenty-one populations and the relative frequency of three different tool types, they found evidence of diversification and cumulative change over time, from simple to complex.

A practical example of the power of social learning comes in the conservation of the California condor by captive breeding and release (Kaplan, 2002). In desperation, as their numbers plummetted, all wild condors were taken into captivity and their offspring hand-reared by human caretakers. When released into nature, these off-spring were hapless, and most died. Their prognosis improved when they were reared with older condor mentors, and success further increased when wild-reared elders and captive-reared youngsters were released together. The elder companions somehow influenced the released birds to use traditional nest sites, water holes and foraging areas.

MAMMALS

Among terrestrial, nonprimate mammals, the best work on social learning is that which combines field and laboratory studies, so that both experimental elegance and ecological validity are intertwined. Terkel (1996) and his students have studied black rats in conifer plantations in Israel. There, the rodents exploit pine seeds, which are extracted by two main techniques: shaving and stripping. Through cross-fostering and observational learning paradigms, intergenerational social learning of food processing has been shown. In nature, the adaptational package adopted by the rats has made them 'ecological squirrels', e.g. they have become secondarily arboreal.

For aquatic mammals, the pinnipeds have long been known for behaviour that appeared culture-based. Sea otters in California are famous for smashing molluscs on anvil stones balanced on their chests (Hall & Schaller, 1964). Their counterparts further up the Pacific coast in Alaska do none of this. A more recent example has harbour seals learning to respond differentially to the cultural differences in the calls of their main predator, the killer whale or orca (Deecke *et al.*, 2002); (see below). The seals respond strongly to the calls of seal-eating whales, and to unfamiliar fish-eating whales, but ignore the calls of familiar fish-eating whales. This seems likely to be social learning because the costs of individual learning (i.e. being eaten!) makes it unlikely that the naïve seal would get a second trial.

CETACEANS

The case for culture in whales and dolphins has been made comprehensively and persuasively by Rendell and Whitehead (2001). What should be emphasised here in a wide-ranging comparison is how hard it is to collect data on cetaceans, especially in the open sea. They roam widely and mostly underwater, sometimes at great depths. They leave few momentary artefacts (e.g. bubble-nets) and, lacking prehensile organs, engage in little complex food-processing. Few questions about cetacean ethology will ever be investigated experimentally, for there is both a practical (large body size) and ethical (high morbidity and mortality) case against keeping cetaceans in captivity. Finally, it is hard for a terrestrial mammal with bipedal locomotion, bimanual manipulation, and limited airborne speech to connect to a virtually limbless aquatic creature with sonar. It is astonishing that we humans can relate to them as well as we do, helped by a combination of our technology and their tolerance.

Inshore dolphins offer the best opportunity for prolonged behavioural observations of known individuals, at least at Shark Bay, Western Australia (Connor & Smolker, 1985). There, bottlenose dolphins in shallow lagoons show some tool-use (Smolker *et al.*, 1997) as well as a fission–fusion form of social organisation (Connor *et al.*, 2000). In this and other ways, they can be said to resemble chimpanzees, yielding a convenient aquatic–terrestrial convergence. However, in addition, the dolphins show behavioural adaptations that exceed all other nonhuman species: they show super-coalitions (i.e. alliances of alliances among males (Conner *et al.*, 1992)) and equally unparalleled vocal communication. For example, not only do they show signature whistles of individuality, but they can match the whistles of their companions in a show of solidarity (Janik, 1999, 2000). The similarities between Shark Bay's dolphins and Gombe's chimpanzees are almost uncanny, even down to the complications of being provisioned by humans. But as hard as it was for Jane Goodall to pursue apes on the steep slopes of the Western Rift in Tanzania, her task was a piece of cake compared with what behavioural dolphinologists have to endure at sea.

Many of the same constraints apply to studies of the largest dolphin – the killer whale – in the northwest Pacific. Again, determination and technology come to the rescue. Two cultural variants coexist in the waters off British Columbia: one hunts for seals and dolphins, in small silent parties that rove widely; the other feasts on salmon close to home, in big, noisy groups. They are the same population genetically, but distinguishable not only by mode of subsistence, but also by vocal dialect (Baird, 2000).

Further south, off the Chilean coast, killer whales offer tantalising evidence of vertical cultural transmission of hunting technique by teaching. There, the killer whales prey upon sea lion pups taken from shingle shores, by temporarily beaching themselves long enough to snatch a pup and then retreat to the water (Guinet & Bouvier, 1995b). It is a high-risk, high-payoff foraging strategy: too far out of the water means fatal stranding on the shore. Thus, teaching by the mother with rescue of her inexperienced youngster is preferable to potentially costly trial-and-error individual learning (Guinet & Bouvier, 1995a).

The most extensive scientific literature on cetacean culture comes from the biggest species – the pelagic whales – on their long-range vocal communication. Extensive reviews can be found in Rendell and Whitehead (2001) and Janik and Slater (1997). A few selected examples here will be enough to show the richness of the singing culture of whales worldwide. Temporal and spatial variation in the courtship songs of the humpback whale is well known (Payne & McVay, 1971). When these broadcasts change rapidly on a transoceanic scale, it is likely that horizontal cultural transmission is responsible (Noad et al., 2000). In another species – the sperm whale – it is the patterns of click sounds (codas) that vary over thousands of kilometres of ocean. Five clans of thousands of whales sharing the same coda are sympatric over the extent of the Pacific Ocean (Rendell & Whitehead, 2003). Finally, cultural selection may explain a puzzling finding in these species of whales with matrilineal social structure. They have reduced mitochondrial DNA diversity compared with nonmatrilineal species. It seems that neutral mitochondrial DNA loci may 'hitchhike' on maternally transmitted strategies of migration, foraging, and babysitting (Whitehead, 1998).

DISCUSSION

What can be said about such a wide range of candidates for cultural status, both in taxa and in behavioural patterns? How can we compare a bird's bower with a whale's song? Also, the literature on purported cultural or culturelike phenomena in nonprimate species varies greatly in terms of context, from tight laboratory experiments (Fiorito & Scotto, 1992) to looser experiments in nature (Diamond, 1988). Observational studies range from completely wild subjects watched opportunistically (Hall & Schaller, 1964) to managed captive colonies (Reznikova, 2001).

Functionally the activities described also vary greatly: subsistence/foraging, courtship, communication, antipredation, locomotion, self-maintenance, social relations, rearing young. However, this variety occurs across species, not within them. Again and again, we seem to confront behavioural adaptation by single species that are good at only one aspect of social learning. The sea otter that shows such flexibility – various invertebrate phyla (e.g. sea urchins, molluscs, crustaceans, etc.) – in extractive foraging by percussive technology shows no such variety in other behavioural realms. The songbird species that are so impressive at rich vocal learning show no similar variation in foraging techniques. This is *not* to say that songbirds lack nonvocal social learning: the famous milk-bottletop-removing tits of the United Kingdom are a well-known example (Fisher & Hinde, 1949). But in no case is one species of bird known for both vocal social learning and visual social learning.

The exception to this one-trick pattern may be the cetaceans: killer whales differ in foraging, communication, and perhaps in greeting and parental behaviour. Humpback whales show a distinctive foraging pattern – lobtail feeding (Weinrich *et al.*, 1992) – as well as their famous long-distance singing (Rendell & Whitehead, 2001). Furthermore, it is only with cetaceans that interactive cultural relations occur between humans and undomesticated nonhuman species. Human–dolphin fishing co-operatives, in which *H. sapiens* and dolphins work together to harvest shoals of fish, exist in West Africa, Brazil, and South Asia (see, for example, Pryor *et al.*, 1990; Simöes-Lopes *et al.*, 1998). These mutually beneficial arrangements by which the dolphins drive the fish into human containment, and then both feast on the trapped prey, appear to have been invented independently several times around the world.

Finally, there is the issue of who transmits knowledge to whom. Culture is collective, but social learning may be dyadic, requiring minimally only a demonstrator and an observer. When a young California sea otter acquires percussive technological skills from his mother there is no evidence that any other otters are involved. Given their solitary lifestyles, it is hard to imagine the sea otters of Monterrey Bay as a collective. Rather than being a culture, it is likely that they have a geographically bounded set of matrilineal traditions of at least one type of food-processing. Scrutiny of the details of the technology might allow this to be tested, if maternally related otters show more resemblance in their techniques, while proximity to nonrelated or paternally related otters shows no correlation to similarity of techniques. Many of the examples of nonprimate candidates for culture show social learning that could be only one-to-one transmission.

Another caveat has to do with the relative novelty of the acts learned. Doing something for the first time that developmentally will become commonplace (e.g. a guppy swimming a particular route) is very different from doing for the first time something that (apparently) has never been done before (e.g. blackbirds incorporating the sounds of touchtone telephones into their songs). It is hard to call the former innovation in any real sense of the word. Of course, cultural patterns need not be inventive, and most are not. Shaking with the right hand in greeting or parting must have been invented at some time, but current practioners neither know nor care. Even less impressive are socially learned acts that show no novelty (i.e. when behavioural patterns already in the repertoire are redeployed in a new context, usually to a new stimulus). Responding with habitual avoidance to a newly recognised threat certainly is adaptive, but it is less impressive culturally than a newer and better response to an old threat.

The main conclusion, however, from all of the nonprimate findings, is that a strong case for social learning is established in many taxa. Apart from the singular case of song-learning in birds, there is little yet that constitutes convincing evidence of culture, as defined above. Behavioural diversity across groups and standardisation within groups is seen, but signs of collectivity and identity are few. All of these features may exist in the cetaceans, but to collect the telling data is cursedly difficult. To put it bluntly, there is nothing in nonprimate social learning, however wonderful and impressive, that could be mistaken for human culture. This is not a book about octopus, bowerbird, or orca. I am a primatologist, and my goal here has been to lay the groundwork for consideration of the prospects and pitfalls of cultural primatology, which begins in the next chapter.

5   Primates

Primatologists seem to assume that monkeys and apes are the most likely nonhuman culture-bearers (McGrew, 1998), while nonprimatologists seem to find this to be presumptuous and exasperating (e.g. Laland & Hoppitt, 2003). Whatever the merits of other mammalian orders (whales and dolphins) or other vertebrate classes (birds), it is probably true that there are more published studies of culture in monkeys and apes than in all other mammalian taxa combined. Cultural primatology is found in both empirical and theoretical articles and book chapters, across several academic disciplines, as discussed in Chapter 3 (e.g. Strier, 2003). In North America, many departments of anthropology, formerly the guardians of the culture concept, have a primatologist, but few anthropologists study any other nonhuman taxa. This means that nonhuman culture as presented in introductory textbooks to anthropology usually is presented solely in primatological terms, which may lead to a biased picture, as discussed in the previous chapter.

Also, the first empirical, as opposed to speculatory, engagement of students of culture and students of animals came in primatology. This began in the summer of 1953, when Imo, a Japanese monkey, was seen to wash her first sweet potato on the beach at Koshima (see below). The resonance of that simple act, and its recognition by Japanese primatologists continues to sound today, a half century later (de Waal, 2001; Matsuzawa, 2003).

Yet in cultural terms, a primate is not a primate is not a primate. The Order Primates contains about 200 species, ± 20 per cent depending on whether you are a splitter or a lumper, but very few forms yield cultural data. I found that a detailed database on culture from the Primate Information Center (PIC) showed 80 per cent of entries to come from only five genera (*Cebus, Gorilla, Macaca, Pan,* and *Pongo*). Of these, over half were from a single species: the chimpanzee (McGrew, 1998). As I write, a similar, more recent bibliography from the PIC on 'Culture, Troop-specific Activities and Custom Spread', yields a similar result: over half the listed references are to the chimpanzee (once abstracts, comments, and redundancies are deleted). The rest of this chapter will concentrate on these taxa, but will leave the chimpanzee until the next chapter.

CAPUCHIN MONKEYS

The genus *Cebus* is a New World radiation of omnivorous, large-brained, prehensile-tailed, and very manipulative primates (Perry & Manson, 2003). Of the four species, the white-faced capuchin monkey has had the lion's share of field work related to culture (Perry *et al.,* 2003b). In captivity, the brown or tufted capuchin monkey has been studied more than all other species of *Cebus* combined (Westergaard, 1994; Fragaszy *et al.,* 2004). This 'mismatch' could have implications for cross-species or cross-context comparisons, or for generalisations about the genus. (See similar points made below for the bonobo and chimpanzee.)

The apparent convergence between New World monkeys and Old World apes has proven heuristically useful, at least for seeing how far the monkeys can match the apes (Westergaard, 1994; Visalberghi & McGrew, 1997). In some cases, studies of capuchin technology in captivity have gone beyond the 'upper limits' of the chimpanzee in taking early hominids as their models, e.g. capuchins can learn to use bamboo tools with cutting edges (Westergaard, 1998).

A parallel between the white-faced capuchin and the chimpanzee that verges on the uncanny is that of hunting for mammalian prey. Rose (1994, 1997) reported social hunting of squirrels that resembles chimpanzees pursuing arboreal monkeys, especially red colobus. The squirrel is chased down by groups of adult monkeys until the prey is surrounded or driven to the ground, then a monkey grabs the prey and dispatches it, and the carcass is torn apart and shared

(Perry & Rose, 1994). The pattern is like a scaled-down version of a red colobus hunt by chimpanzees (Stanford, 1998a), but there are inter-populational differences in the details: one group of capuchins uses an efficient cranial killing bite while another starts eating the prey while it still lives (Perry et al., 2003b). Like chimpanzees who steal bush piglets from terrestrial nests by goading the adults to attack, so capuchins rush in to snatch the pups of coatis from their nests (Rose, 1997).

The extent of elementary technology in captive brown capuchins is extraordinary, exceeding that of all other species of non-human primates (Westergaard, 1994, 1995; Westergaard & Suomi, 1995). However, none of the habitual tool use was spontaneous; rather, carefully scaffolded problems were set to a test-savvy group of monkeys whose previous efforts on other problems had led to rewards. These were a small set of canny problem-solvers posed with a series of experiments featuring the tool of the month. The obvious question is that of ecological validity; few of these contrived situations inducing tool use have counterparts in the real world of neotropical forests.

Wild white-faced capuchins have a repertoire of motor patterns for object manipulation, especially in food-processing. They pound, rub, tap, and wrap food items to prepare them for consumption (Panger et al., 2002b). Some of these are tool use and even tool-making by Beck's (1980) definition. Similar findings have been reported for wild brown capuchins (Boinski et al., 2000). Even more impressive are single-population reports of wild capuchin use of a weapon (Boinski, 1988), hammer and anvil (Fernandes, 1991; Oxford, 2003), leaf contain-ers (Phillips, 1998), and probes (Chevalier-Skolnikoff, 1990). At first glance, the elementary technology of the capuchin radiation seems equivalent to that of the chimpanzee.

Like the published literature on chimpanzee material culture, that of capuchin monkeys varies from long-term studies lasting years to natural history notes. Unlike chimpanzees, where the study sites across Africa span thousands of kilometres from Uganda to Senegal, the four key sites in Costa Rica for white-faced capuchins are close together: Palo Verde, Lomas Barbudal, Santa Rosa, and Curú were all connected by contiguous forest until as recently as 50 years ago (Perry *et al.*, 2003b). This makes the cultural diversity that has emerged in such a short time all the more impressive.

So far, almost all cultural traits proposed and discussed for capuchin monkeys have to do with subsistence, in foraging and food-processing (see, for example, Rose, 1994). This can be interpreted by sceptics as behavioural diversity that reflects only environmental constraints and affordances, and so always in principle can be explained by individual learning in transaction with the environment, rather than by social transmission processes (Kummer, 1971). For example, some groups of white-faced capuchins regularly follow army ants on the forest floor, in order to catch insects scared up by them, while other groups of capuchins in similar forests totally ignore this opportunity (Perry *et al.*, 2003b). It is always possible that the ants are slightly different, or the substrate, or the fleeing insects, etc., and if so, a cultural explanation is not needed.

Social customs or conventions (apparently the terms are synonymous) cannot be so dismissed, especially if they are arbitrary in form, as with the grooming hand-clasp of chimpanzees (McGrew & Tutin, 1978; de Waal & Seres, 1997; Nakamura & Uehara, 2004). These customs do exist in capuchins: Perry *et al.* (2003a) define a social convention as 'dyadic social behaviours of a communicative nature that are shared among members of particular social networks'. Their three candidates for social traditions are: (1) hand-sniffing (placing fingers in or over another's nose), (2) sucking bodyparts (such as another's finger, toe, ear, or tail), and (3) games (tugging and biting at another's finger, hair, or a detached object). In the latter game, the object essentially becomes a toy. Such conventions may originate and flower with one keen practioner, and then die out accordingly, but they may last for years, and so qualify as true traditions.

As recently as 5 years ago (McGrew, 1998), the cultural primatology of capuchin monkeys was at the ethnographic stage. Now, exciting ethnology is being done (Perry & Manson, 2003). The only thing left that keeps chimpanzees ahead of capuchins in the culture game is the head start that the researchers on apes had in field work, with studies that began in the 1960s and 1970s for *Pan* versus those in the 1980s and 1990s for *Cebus*.

MACAQUE MONKEYS

The evolutionary radiation of the macaques remains one of the most widespread and successful, with a distribution from Japan across Asia to the Mahgreb. One species of macaque – the rhesus monkey – has colonised successfully the New World – in Puerto Rico (Rawlins & Kessler, 1986) and Florida (Maples *et al.*, 1976b) – with human assistance at the outset. Another – the long-tailed macaque – has colonised islands in the Indian Ocean (Mauritius: Kondo *et al.*, 1993) and Pacific (Palau: Kawamoto *et al.*, 1988). Macaques occur from mangrove swamps to Himalayan foothills, and are the only primates adapted to snowy winters – in northern Japan and in the Atlas Mountains of Morocco. Therefore, it comes as no surprise that these Old World monkeys show the behavioural flexibility that is a prerequisite to culture.

An unusual example of behavioural variation in the rhesus macaque is a form of stone-handling called 'stone-stuffing' (Clarke *et al.*, 1993). It seems to occur in only one captive colony: the Tulane Primate Center in Louisiana, United States. Some adult female monkeys fill their cheek pouches with pebbles, sometimes to the extreme extent of producing gross bulges that could be mistaken for advanced goitres. The stones are retained in the pouches and become polished from continual friction (see McGrew, 1992, p. 203, for references to gastroliths in sea lions). What is the reason for this bizarre habit? It is probably not nutritional, given the type of stone, but may function in dental hygiene, as stone-stuffers' teeth were in better condition than matched controls of similar age, sex, and rank. Is it cultural? Only a follow-up study will tell, to see if the habit has spread horizontally or been passed on vertically to the females' offspring.

The most complete ethnographic record of any species of macaque is that of the Japanese monkey. It has the widest latitudinal distribution of any nonhuman primate, from the tropics of the Ryukyu Islands to the ski slopes of northern Honshu. It is the only species of primate to coexist with humans in a modern, industrial society. But the species' position as the subjects of pioneering cultural primatology is due to one visionary primatologist, Kinji Imanishi, who with two students, Itani and Kawamura, began in 1948 studies of Japanese macaques on Koshima, an offshore island (de Waal 2001; Matsuzawa, 2003). From this research emerged fundamental findings on kinship and social rank, but also on acquired and propagated new behavioural patterns: sweet-potato washing, wheat-sluicing, sea-bathing, and catching thrown food items (Kawamura, 1959; Kawai, 1965). The transmission, both horizontally and vertically, of the novel behavioural patterns has been monitored from the outset; however, diffusion across troops could not be studied, given the isolated population.

In sweet-potato washing, dirty tubers were dumped by the researchers on the beach for the monkeys to eat, and Imo, a 1.5-year-old female, invented a technique of cleaning the potatoes by washing them in the stream nearby. The food-processing technique spread to her peers, their siblings and their mothers, but not to adult males, who ignored these (childish?) acts. As the spud-scrubbers matured and had offspring, the mothers passed on the habit to their youngsters, in orthodox downward vertical transmission.

Similarly, 4 years later, Imo invented wheat sluicing, in which mixed handfuls of wheat and sand were thrown into the sea. These cereals, too, had been provided by the researchers. The sand sank, the grain floated, and the wheat could easily be skimmed off the surface; this was much more efficient and less gritty than picking up grains of wheat one by one from the beach.

In both cases, the inventive behavioural patterns evolved toward greater efficiency and comfort, and yielded unexpected spin-offs. Sweet-potato washing progressed from fresh to salt water and from the surf to excavated small pools (Watanabe, 1994). Wheat-sluicing graduated to hand-held (as opposed to thrown) grain, which was easier to defend against pirating companions (Kawai *et al.*, 1992). From familiarity with the sea came wading, then swimming, then underwater foraging for marine prey, e.g. octopus and limpet (Watanabe, 1989).

Why were the Koshima macaques so enterprising? It seems to be another case of necessity as the mother of invention, with a little help from humans. Koshima is small and sparsely resourced; once the provisioning plus innovation took off, infant mortality dropped and the population numbers swelled (Watanabe, 2001; Matsuzawa, 2003). But such innovation was not confined to Koshima. Similar cultural progress occurred in the Shiga Heights after Suzuki (1965) induced the snow monkeys to enter the natural hot springs near Nagano. He threw apples into pools to introduce the monkeys to the novel environment, and they came to exploit this for winter thermoregulation. In a long, hard, cold winter of subsisting on the cambium layer of conifer bark, gaining precious energy savings by prolonged soaking proved to be advantageous.

As some critics suggest that Japanese monkeys only show such habits when provisioned by humans, some natural behavioural patterns should be mentioned. Wild monkeys living far from the sea wash natural and dirty roots in a stream (Nakamichi *et al.*, 1998), and the monkeys of Yakushima specialise in preying on frogs and lizards, unlike any other populations of Japanese macaques (Suzuki *et al.*, 1990). Yakushima's monkeys live in a large national park and have never been provisioned. McGrew (1998) gives other examples.

Finally, lest it be said that all traditions of Japanese monkeys have to do with subsistence (food-processing) or self-maintenance (thermoregulation), there is stone-handling. First described in a single troop at Arashiyama near Kyoto 25 years ago (Huffman, 1984), this behavioural pattern is known to exist in at least five other isolated populations across Japan. It is a complex of eleven motor patterns, all to do with handling pebbles: pick up, pile, scatter, roll, rub, clack, carry, cuddle, drop, toss, push (Huffman, 1996). None of these is functional, i.e. no overt goal is attained. It is common – several times per hour (personal observation) – and widespread – all infants acquire it by 6 months of age. Why do they do it? It may be that the stones function as worry beads.

Western comparative psychologists and other critics continue to be dismissive of the cultural achievements of Japanese macaques (Galef, 1992; Boyd & Richerson, 1996; Tomasello & Call, 1997, pp. 276–8; Tomasello, 1999; Laland & Hoppitt, 2003; van Bergen et al., 2003). Most of these commentators are ill-informed from either relying on biased secondary sources (often each other) or choosing to read only the early papers of the 1950s and 1960s, instead of the updates written from the 1990s onward (see references above).

What do their criticisms amount to? One point made is that the behavioural patterns employed are simple ones common to all macaques. This is true of some, but so what? This is like saying that entering data on a keyboard is of no interest because it is the same motor pattern as tapping your fingers in boredom. A second criticism is that the behavioural innovations spread too slowly. When 'too slowly' is explained, it turns out to be relative to a simple-minded model of equiprobability of transmission to all group members. This is nonsensical, since social interaction in macaques is not random or uniform, but highly structured according to rank, kinship, and age. It is not surprising that sweet-potato washing did not spread willy-nilly, like gas molecules bouncing around in a chamber.

The third and most often repeated criticism invokes the 'ratchet effect' hypothesis. It poses the possibility that sweet-potato washing is not learned socially, but instead is learned individually, so that each monkey 'reinvented the wheel' (Tomasello & Call, 1997, p. 276; van Bergen et al., 2003). A slightly more generous alternative has the naïve monkey learning through local enhancement (Boyd & Richerson, 1996). The effect is the same, for in this interpretation each new monkey must start from scratch, and so no individual can profit from the improved performance of his predecessors. Even if a gifted monkey devises an improved technique of washing, this memetic mutation will go nowhere, says the argument. In contrast, creatures that have observational – especially imitative – learning, can have cumulative cultural change, i.e. beneficial modifications to existing habits can accumulate over time. In short, progress. Thus, humans can make history, but not other creatures. (Why this is called a ratchet is another question. Ratchets move only in one direction, and clearly human culture and probably nonhuman culture can decline as well as progress.)

There are two main problems with the ratchet hypothesis. One is practical, as outlined earlier: sweet-potato washing and wheat-sluicing *have* elaborated and become more profitable and convenient over time. But even in principle, local enhancement can provide a basis for cumulative change. When the monkeys made the change from fresh water to tastier salt water with their spuds, some individual had to be the first to do so, and that potato-washer then enhanced the stimulus value of sea-water. If anyone paid attention, then the number of washers using salt-water increased. For wheat-sluicing, the innovator of pool-digging therefore enhanced a pool of water on the beach, whereas waders into the surf had before enhanced only the waves. Local or stimulus enhancement *is* social learning (Whiten & Ham, 1992) and though it may not be as efficient as more complex forms, it can drive cultural change, as more than 50 years of data from Koshima seem to show us.

GREAT APES

On phylogenetic grounds, one would expect to find rich and varied cultures in the great apes in nature. They are large-brained creatures who show comparable levels of high intelligence when tested in captivity (Byrne, 1995). All four species – bonobo, gorilla, orangutan, and chimpanzee – show similar problem-solving abilities and facility at acquiring human systems of communication, either by gesture or computer. All four species show similar ability at mirror-image recognition (Inoue-Nakamura, 1997), a cognitive capacity that correlates with tool use (McGrew, 1989). Yet, only the chimpanzee shows con-vincing evidence of culture on both material (McGrew, 1989) and social fronts (cf. Hohmann & Fruth, 2003). For each of the other three species, there are special issues, usually obstacles.

The bonobo is very close to the chimpanzee genetically and ecologically, yet seems so far away culturally, in terms of our limited knowledge. But do we know enough yet to draw any conclusions? All wild bonobos live in one country – the Democratic Republic of Congo (DRC) – a homeland racked by internal strife. Consequently, as I write, all the long-term field sites are inactive, and only two of them – Lomako and Wamba – had habituated subjects before the shutdown (Thompson, 2001). So, what we know of bonobo culture may be only a fraction of what there is to know, and only time and political stability will tell (Stanford, 1998b).

An apparent exception to this frustrating ignorance is material culture. Even totally unhabituated apes leave behind remnants and artefacts, and if these were there, keen-eyed primatologists would have found them, but they have not (cf. Bermejo et al., 1989). If bono-bos were cracking nuts, fishing for termites, or dipping for driver ants, we should know it from the tools on the forest floor. There is yet no evidence of elementary technology being used in subsistence or as weapons, yet these patterns are universal in their sibling species: the chimpanzee (Ingmanson, 1996; Whiten et al., 1999). The best that can be said for Wamba is that bonobos occasionally use leaves as protec-tion from the rain, either as hats or umbrellas (Ingmanson, 1996).

To make a start in clarifying the cultural position of the bonobo, Hohmann and Fruth (2003) have catalogued candidate behavioural patterns from Lomako. They found that fourteen of the sixty-five cat-egories of cultural candidacy listed by Whiten et al. (1999) for chim-panzees also occur in bonobos. They listed another nine behavioural patterns in Lomako's bonobos that have not been reported for the chimpanzee, including wading in foraging, covering the body with leaves in bed, using the feet to detach large fruits, and chasing duikers.

Some of these differ from Wamba, the only other bonobo field site with sufficient ethnography to allow comparison (Nishida *et al.*, 1999). Lomako's bonobos subdue, kill, and eat duikers, then share the meat. Wamba's bonobos ignore duikers, although they eat small mammals (Ihobe, 1992). Some behavioural patterns (e.g. drum and branch-drag) appear to be bonobo universals, just as they are in the chimpanzee. Foraging in streams and pools is widespread across bonobo populations, but there is postural variation: in the shallow waters of Lomako, the apes are quadrupedal, but in the deeper waters of Wamba and Lukuru, the apes go bipedal (Hohmann & Fruth, 2003). Most (thirteen of twenty-three) of the candidate patterns for culture at Lomako function in social life, especially in communication.

For the other African ape – the gorilla – the story is similar. The vast majority of the species are western lowland gorillas widely distributed about the Congo river basin and further north to the Cross river area of Nigeria. Despite excellent long-term ecological study at sites in Gabon, Republic of Congo (Brazzaville), Central African Republic, and Cameroon, little is known of their behaviour. There is no habituated population of western lowland gorillas, and the only behavioural data come from open marshes (*bai*) where gorillas occasionally emerge into the open to eat aquatic vegetation. Even in this brief window, there are tantalising glimpses: at Mbeli Bai in Nouabalé-Ndoki National Park, Republic of Congo, adult males seek to intimidate rivals with a splash display (Parnell & Buchanan-Smith, 2001).

Eastern (lowland) and mountain gorillas at higher altitudes in the Democratic Republic of Congo, Rwanda, and Uganda suffer from the same handicaps as bonobos. They are unlucky to be living in war or refugee zones. Many of the habituated eastern gorillas of Kahuzi-Biega National Park, DRC, died in the civil war or from its side-effects. The best-known gorilla population in the world – that of the mountain gorillas of the Virunga volcanoes (Fossey, 1983) – has survived fierce fighting and invasion of its range by refugees and their livestock, but research remains minimal by prewar standards (Williamson, personal communication). Given the extremes of its high-altitude environment (as the only wild-ape population to experience snowfall), it is hard to compare these apes' unique behavioural features to those of other gorilla populations. Virunga gorillas are the only population known to eat driver ants, which are common throughout this ape species' range (Watts, 1989), but they do so without tools, unlike chimpanzees (McGrew, 1974). Is this insectivory a clever innovation by mountain gorillas, or is it driven by their highly marginal habitat? Some clarification could come from the other mountain gorilla population at Bwindi, Uganda, but these apes, like all the others, are yet to be habituated to human observation.

There are other tempting candidates for cultural variation in gorillas: Western lowland gorillas live in classic one-male groups, i.e. one silverback plus several females and their offspring (Robbins, 2001). Mountain gorillas often live in multimale groups with as many as five silverbacks. The latter groups can have 4 times as many members on average as the former. So, is there a real difference in social structure? If so, does this reflect the multiple ecological differences between alpine meadows and lowland rainforest? Or are the bigger mountain gorilla groups just unfissioned multiples of one-male groups unnaturally kept together by the threats from human incursion?

Because the gorilla has even less elementary technology than the bonobo, tackling culture in this great-ape species is an even greater challenge. There are no convenient tools with which to make an analytic start, while habituation proceeds. But we know enough from ethological studies of the mountain gorillas in the Virungas to say that the potential is high for cultural variation in the food-processing of plant foods. Byrne and Byrne's (1993) elegant studies of the sequence of handling of four species of herbs shows evidence of programme-level imitation (Byrne, 1995).

Our knowledge of culture in the Asian great ape – the orangutan – has improved in recent years, but the findings are bittersweet. Deliberate (for farming and logging) and inadvertent (forest fires) deforestation has caused notable loss of orangutans in both Borneo and Sumatra. Despite decades of study in Borneo, the first evidence for habitual elementary technology in orangutans appeared only in the mid 1990s at Suaq Balimbing in Sumatra (van Schaik *et al.*, 1996). (Tool use by captive orangutans is well known, whether elicited or spontaneous (see, for example, Lethmate, 1982; O'Malley & McGrew, 1999).) All earlier reports were anecdotal or marginal (Galdikas, 1982, 1989). But just as the Suaq Balimbing apes showed their technological colours, the population was extinguished by forest clearance, so we will never be able to study further their material culture.

Meanwhile, a coalition of nine field workers studying wild orangutans at six sites – four in Borneo, two in Sumatra – compared systematically their data on behavioural patterns suspected of being cultural (van Schaik *et al.*, 2003). They presented thirty-six patterns: of these, twelve were too rare to analyse and five had likely ecological explanations. The remaining nineteen ranged from using leaves to make a sound (Kiss-squeak with leaves) to taking shelter from rain under a nest (Hide under nest) to using leaves as 'gloves' to handle spiny fruits (Leaf gloves).

One can divide the nineteen behavioural patterns by function as: *subsistence* (i.e. finding, accessing, and processing food and drink); *social* (i.e. interactions with conspecifics), or *self-maintenance*, all else, typically being activities that enable or facilitate individual survival (McGrew *et al.*, 2003). For the orangutan populations combined, most ($n = 17$) were Maintenance, with Subsistence ($n = 12$) and Social ($n = 7$) patterns being notably fewer. This functional profile differs significantly from the comparable breakdown by function for chimpanzees: Subsistence ($n = 34$), Social ($n = 17$), Maintenance ($n = 14$), (Whiten *et al.*, 1999, 2001). Given the solitary inclinations of orangutans, it is surprising to see that social patterns form a similar proportion of their cultural repertoire as for the gregarious chimpanzee (orangutans 19 per cent versus chimpanzees 26 per cent). The difference arises in chimpanzees showing more cultural variation in subsistence (52 per cent) than do orangutans (33 per cent), while the reverse is true for maintenance (orangutans 47 per cent versus chimpanzees

22 per cent). Seven of the seventeen maintenance patterns for orangutans focus on their sleeping nests; unfortunately, a similar degree of attention has not been focused on nest-related activities in other apes (cf. Fruth & Hohmann, 1996).

A closer look at a particular pattern – the use of sticks or twigs to dislodge the seeds of *Neesia* fruits – is informative (van Schaik, 2003). This is an issue of both efficiency and comfort, as the fruits are large and woody and the seeds embedded in massed stinging hairs. It is not necessary to use tools, just better; some keen orangutan populations in Borneo harvest the seeds by hand (van Schaik & Knott, 2001). The prize is worth the effort, as the seed is high in energy. The use of tools to make food-processing easier is found in several places in Sumatra, including unhabituated orangutans who drop their tools under *Neesia* trees, to be found later by field workers. The parallels with the oil palm nut-cracking of chimpanzees (see Chapter 7) are notable: disjunct cultural distribution, skillful task, high payoff goal, etc.

The cultural activities of the orangutans of Suaq Balimbing are not limited to subsistence technology. They also have a coincident gesture and vocalisation at the time of nest-making, which involves passing a leafy twig by the mouth, while making a spluttering sound. The behavioural pattern seems functionless – unless they were saying 'Good night' to their companions – but is shown by all well-observed individuals in the population (van Schaik, 2003).

So, why do the orangutans of Suaq Balimbing have such a rich social life? First, they show the highest population density of all known wild orangutans, at about $7/km^2$ (van Schaik, 2003). This density is higher than that of most wild chimpanzee populations. At savanna sites in Senegal, the density was $0.1/km^2$ (Pruetz *et al.*, 2002); at rainforest sites in Ivory Coast, the average density was $1.7/km^2$ (Marchesi *et al.*, 1995). Second, the Suaq Balimbing apes travel and forage in parties, not solitarily (van Schaik, 1999). Therefore, the opportunities for social learning are much greater than elsewhere. Finally, the Suaq Balimbing orangutans commonly share food among adults, an indication of congenial social relations. All of this adds up to a prevailing atmosphere of social tolerance (van Schaik, 2003).

DISCUSSION

It is hard to generalise about behavioural variation that may be cultural in primates. Some of the least interesting problems have to do with methodology: take, for example, the simple issue of frequency. There is a world of difference between an anecdote and a custom. An *anecdote* is a rare or even single record of an event, while a *custom* is indicated when all able-bodied members of the appropriate age–sex–rank class show the behaviour in regular, predictable, fashion.

Thus, the claim that wild white-faced capuchin monkeys use clubs as weapons is based on a single event, when monkeys attacked a venomous snake with a stick (Boinski, 1988). Similarly, the claim that wild brown capuchins use hammer and anvil to crack open molluscs is based on one monkey using one tool on one occasion (Fernandes, 1991). These are wonderful natural history notes but are hardly the basis for generalisation about a species, or even a genus. Yet this is exactly what secondary (review articles) and tertiary (textbooks) sources do. Sarringhaus (unpublished data) has shown that over two-thirds of citations of anecdotes in primatology are incorrect, usually through overgeneralisation.

Why are anecdotes so problematic? Simply because, of the six possible explanations for any anecdotal event, five are wrong. Here is a telling example. We recovered a knotted 'necklace' of red colobus skin worn by an adult female chimpanzee, Akko, at Mahale (McGrew & Marchant, 1998), (see Figure 5.1). Was this the first record of a manufactured ornament in a wild ape?

Perhaps, but not likely. It could have been: (1) accidental, from Akko's repeated manipulation of the skin, so that a knot got tied inadvertently; (2) mistaken, as Akko might have been trying to bandage her cut finger with the skin but instead tied a knot; (3) observer error, as we might have mistaken Akko's draping a piece of bark around her neck for her wearing of the skin, (4) misattribution, as

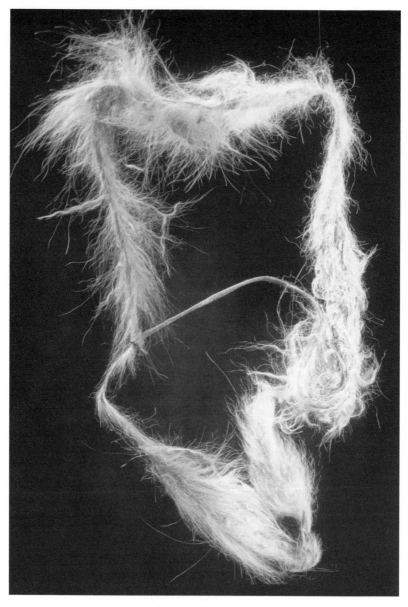

FIGURE 5.1 Knotted skin of red colobus monkey, worn as necklace by adult female chimpanzee at Mahale. Knot is visible halfway down left side.

the knot might have been tied by a baboon, and then only found by Akko; (5) hoax, as our field assistant might have knotted the skin, as a joke. Or the one useful explanation could be that this was (6) the invention of the necklace in Mahale's M-group, to be followed perhaps by a whole fashion for body ornamentation. Sadly, it apparently was not the start of a fad; since recovering Akko's knot in 1996, we know of no further instances. It may have been one of those many memetic mutations that never caught on.

But, you may say, the anecdote of Akko's knot at least shows that chimpanzees have the capacity to use a knot to make a necklace. Wrong. An accident, mistake, error, etc., says nothing about an ape's capacity to do anything. All that an anecdote can do is alert us to a possibility, so that we will look for it again. And this alone is important enough for anecdotes to be published, to draw others' attention to un-expected events. The only thing to do with $n = 1$ is to seek more data.

At the other extreme are customs: well-established behavioural patterns performed appropriately, i.e. by that able-bodied proportion of the group of the right sex, age, rank, kinship, experience, etc., in the right context. These dependent variables as constraints might seem obvious, but some critics of cultural primatology have called for una-nimity, i.e. refusal to grant cultural status unless all members of the group show the behavioural pattern (Tomasello, et al., 1993). This stricture is unrealistic. Only females suckle, so variants in weaning behaviour will not be shown by males. Only independent males go on territorial patrol, so babes in arms will not show this pattern. Only subordinates pant-grunt in submission, so the alpha male will never do so, until he is deposed. And so on. Further, we can hardly expect crip-pled individuals to perform complex manipulations (Stokes & Byrne, 2001), although their ability to adjust to handicap can be remarkable.

In between the extremes of anecdote and custom are woollier intermediary states. A behavioural pattern may be done repeatedly, but only by one individual. Such idiosyncrasy in itself says nothing about culture as a social process, but when that individual influences others, it can be important. Imo, the female Japanese macaque, invented both sweet-potato washing and wheat-sluicing. The remarkable Guapo, a male white-faced capuchin monkey, invented all three games seen at Lomas Barbudal, though he ceased to play them after becoming alpha male (Perry *et al.*, 2003a, b).

Van Schaik *et al.* (2003) showed three categories of intermediate frequency status: *habitual*, when shown by at least several relevant individuals; *rare*, which was undefined but presumably uncommon, and *present*, of unknown frequency but probably rare (see Whiten *et al.*, 1999, for a similar scheme). Such gradations in frequency may be real, or they may reflect incomplete habituation of subjects. True frequencies of any behavioural pattern can be got only from subjects amenable to unlimited focal-subject sampling; anything less is open to bias, e.g. intimate acts such as birth have to be customary but are seldom seen!

Everything said so far about the frequency of cultural traits depends on having behavioural data. But, what about unhabituated populations? Some authors have tried to rate frequency based on numbers of artefacts (McGrew *et al.*, 2003), on the ground that hundreds of tools at tens of worksites should mean more than a few tools at one site. However, such indirect measures of frequency can never show customary status, for even hundreds of tools may have been used by only a few, or even one keen individual.

All of the taxa highlighted in this chapter show customs: white-faced capuchins show forty differences in food-processing technique across three sites, and nine of these are customary (Perry *et al.*, 2003b). For Japanese macaques, customary status in sweet-potato washing is documented in recent analyses (Hirata *et al.*, 2001a; Matsuzawa, 2003). For bonobos, five behavioural patterns are customary at Lomako: branch-drag, branch-clasp, branch-shake, branch-slap, and buttress-beat (Hohmann & Fruth, 2003). For orangutans, fourteen of the nineteen categories classed as 'very likely cultural variants' are customary, and these customs are scattered over all six field sites (van Schaik *et al.*, 2003).

On another front, none of the above species of primates is a 'one-trick pony', in terms of the number and breadth of cultural patterns detailed above. Furthermore, all four species show apparent cultural variation across populations in social, subsistence, and maintenance activities. Finally, at least three of four species show indirect evidence of social learning, e.g. structured spread of behavioural patterns that reflects social relations.

For all other primate taxa, the results are equivocal, for one reason or another. For gorillas, a systematic comparison remains to be published, but lacking habituated subjects in almost all field sites makes any comparative analysis difficult. For other species that have been studied at multiple sites for decades (e.g. baboons), it remains for researchers to collaborate on joint analysis of pooled datasets. For other *Cebus* species – especially the brown capuchin – recent studies show similarities, but also differences, to the white-faced capuchin. (Boinski *et al.*, 2001). This makes generalising at the generic level risky, just as it is for *Pan* (see below).

But is any of this culture? The answer, as always, depends on definition. Table 5.1 is a matrix of four populations of different species (columns) versus eight conditions for culture (rows), with the latter based on Kroeber (1928), as modified by McGrew and Tutin (1978). Notably, the two species of monkeys fulfil more conditions than the two species of apes, but this probably is related to the shorter duration of data collection from the apes, and to their prolonged life-spans. It is harder to look at tradition in terms of intergenerational transmission when a species does not reproduce until its second decade of life! Note that diffusion is absent from all four populations, as none has recorded a behavioural pattern spread from one group to another.

According to van Schaik (2003), a local variant in behaviour can be considered as a tradition if it meets three criteria: (1) *common*, shown by several individuals (habitual, but not necessarily custom-ary, in the terms given above); (2) *long-lasting*, persisting across generations, and (3) *social learning*, maintained by it in some form. The third condition can never be fulfilled in nature by experimentalist

Table 5.1 *Conditions for culture, as modified from Kroeber (1928; see also McGrew & Tutin, 1978) as seen in four species of nonhuman primate. Key: + = present; blanks = unknown.*

	Species (and site)			
Condition	White-faced capuchin (Lomas Barbudal)	Japanese monkey (Koshima)	Orangutan (Suaq Balimbing)	Bonobo (Lomako)
Innovate	+	+		
Disseminate	+	+		
Standardise	+	+	+	+
Durable	+	+		
Tradition		+		
Transcendant Diffusion	+	+	+	+
Natural	+	+	+	+

standards (see Chapter 4), but can be inferred by indirect means. Table 5.2 casts his three criteria across the same four species, and the results are similar to those in Table 5.1, except that there are not yet enough data from Lomako to confirm social learning in bonobos.

For Perry *et al.* (2003b), seven criteria must be met before a practice can be considered a tradition. The practice must: (1) show intergroup variation; (2) occur at least once per 100 observation hours; (3) be performed by at least three individuals; (4) show intragroup variation; (5) increase in number of performers over time; (6) last for at least 6 months, and (7) be aided in transmission by social context.

Table 5.3 presents these seven criteria in terms of the four species. The results are also similar, and again the apes show blank cells in the matrix through lack of observations on fully habituated individuals over a long-enough timespan.

A very similar result emerges if we use the working definition proposed by nonprimatological sceptics, as aired in Chapter 4

Table 5.2 *Criteria for tradition, taken from van Schaik (2003) as seen in four species of nonhuman primate.*

Criteria	Capuchin	Macaque	Orangutan	Bonobo
Common	+	+	+	+
Enduring	+	+	+	
Socially learned		+		

Table 5.3 *Criteria for tradition, taken from Perry & Manson (2003), as seen in four species of nonhuman primate.*

Criterion	Capuchin	Macaque	Orangutan	Bonobo
Intergroup	+	+	+	+
Frequency	+	+	+	+
Common	+	+	+	+
Intragroup	+	+	+	
Practitioners	+	+		
Enduring	+	+	+	+
Social context	+	+	+	

(Laland & Hoppitt, 2003). They required candidate behavioural patterns to show five elements: (1) *group-typical* (shown by majority of group?); (2) *shared* (same behavioural variant shown by interactants?); (3) *community members* (individuals show signs of group solidarity?); (4) *social learning* (individual's behaviour changed by exposure to others?), and (5) *social information transmission* (individual's mind changed by exposure to others?). The parenthetical phrases ending in question marks are my attempt to explain the elements, as there is none in the original text. Table 5.4 gives the breakdown of the four candidate species on these five criteria. Again, the results are similar, though this definition fails to discriminate between the monkeys and the apes. Orangutans are not granted community membership on grounds of their solitary social structure, but for at least the Suaq

Table 5.4 *Characteristics of nonhuman culture (taken from Laland & Hoppitt, 2003) as seen in four species of nonhuman primate. Key: ? = unable to test.*

Characteristic	Capuchin	Macaque	Orangutan	Bonobo
Group-typical	+	+	+	+
Shared	+	+	+	+
Community members	+	+		+
Social learning	+	+	+	+
Social information	?	?	?	?

Table 5.5 *Characteristics of culture, as 'the way we do things' (Chapter 2), as seen in four species of nonhuman primate.*

Characteristic	Capuchin	Macaque	Orangutan	Bonobo
Do things (behaviour)	+	+	+	+
The way (standardised)	+	+	+	+
We (collectivity)	+			
Way we do things (identity)				

Balimbing apes there may be enough social interaction to consider them a 'spaced-out' community (van Schaik, 1999).

Finally, what of my simple definition (in Chapter 2) of 'the way we do things'? Standardised behavioural patterns are present in all four species (see Table 5.5), but only the 'games' of the white-faced capuchin appear as meaningful interactions of social significance, and so indicative of collectivity. For the bonobo, several patterns (e.g. Leaf-clip and Groom-slap (Hohmann & Fruth, 2003)) seem to be good candidates for collectivity, but the data need to go beyond description to analysis before they can be considered convincing. For the orangutan, it is asking a lot to seek for collectivity in such a widely spaced species, but a behavioural pattern such as mutual genital rubbing by females, which rarely is seen in three populations but is apparently absent in three others (van Schaik *et al.*, 2003), has promise. More data are needed.

For exploring issues of identity – the epitome of 'the way we do things' – none of the four species yet present data on either emigration or immigration by known individuals. Only for Lomako's bonobos are data given for intergroup encounters, and these are *Pan*-typical agonistic display patterns using the living branches of trees.

So, is there culture in nonhuman primates, and if so, can we therefore do cultural primatology? Yes, according to some definitions; no, according to others. You pays your money and you takes your choice. Chapter 6 tackles the same issues with one more species – the chimpanzee – to see if the picture is any clearer for that African ape.

6 Chimpanzee ethnography

Ethnography has been the heartbeat of the anthropological endeavour since the emergence of the discipline in the late nineteenth century. It is a set of methods by which anthropologists describe systematically, and record, the rich cultural tapestry of humanity. There was a time when the budding sociocultural anthropologist had to go to the field and then bring back an ethnography (i.e. a comprehensive account of a traditional society), the more exotic the better. Indeed, the great names of anthropology – Malinowski, Radcliffe-Brown, Mead, etc. – made their reputations in just this way. (Ironically, by the end of the twentieth century, these pioneering efforts were being shunned and disowned for their supposed imperialist, racist, sexist, etc., biases (see Aunger 1995, for an example of the debate).)

The composite of ethnographies for *Homo sapiens* worldwide is the ethnographic record. It is the master database for cross-cultural comparisons that yield both human universals and variation. The extent of information is such that massive archives are required, e.g. the Human Relations Area File (Murdock *et al.*, 1965), and extensive atlases are needed to access them (see, for example, Price, 1990). Since the body of knowledge is so great, coding the topics of content (e.g. cooking, rites of passage, etc.) allows for wide-scale, secondhand analyses without ploughing through thousands of pages of text (see, for example, Murdock & Morrow, 1970). From the ethnographic record came studies qualitative or quantitative, descriptive or hypothesis-driven, impressionistic or reductionist. This is ethnology, and it constitutes the principle justification for the existence of anthropology as a discipline. No other branch of the social sciences or humanities has a wider scope: Price's (1990) atlas gives information on more than 3500 cultural groups.

If ethnography is essential to the study of human culture, then cultural primatologists should pay attention. Before tackling this prospect, however, an earlier stage of culturology must be addressed: natural history. Early students of human cultural variation were interested equally in plants, animals, and primitive humans. Explorers and travellers sought specimens of all of the above. Plants were pressed, skins cured, and artefacts bartered. (Sometimes the latter objects and their owners were got by less honourable means, alive or dead (Thomas, 2000).) The material proceeds of natural history often ended up in museums of the same name, e.g. London's Natural History Museum or New York's American Museum of Natural History. From these institutions, and later from universities, came ethnographers. Today, the juxtaposition between nonhuman nature and human culture side-by-side in the same exhibition is considered by some to be politically incorrect. But historically, just as natural history preceded ecology in the natural sciences, so natural history set the stage for ethnography in the social sciences.

So, can ethnography be done on nonhuman species? Answers to this question vary greatly: of Wrangham *et al.* (1994), Ingold (2001) has said that the primatologists '. . . do not know what they are talking about', (p. 337). To Ingold, ethnography should '. . . achieve an *understanding* that is sensitive to the *intentions* and *purposes* of the people themselves, to their *values* and *orientations*, to their ways of *perceiving*, *remembering*, and *organising* their *experience*, and to the *contexts* in which they act,' [italics are mine]. Thus, proper ethnography goes beyond the '. . . mere recording of observed behaviour' to at least ten types of inference (italicised above). Ingold welcomes such an undertaking by those who study nonhumans, but insists that they eschew positivism and reductionism in doing so. The issue thus becomes one of inference and its limits, and researchers in animal cognition investigate all ten of the phenomena laid out above. Most of their empirical work is done experimentally with captive subjects, but as long ago as 1971, Goodall presented an ethnographic account of wild chimpanzees that addressed all ten of Ingold's topics.

Contrary to Ingold, students of nonhuman species (e.g. McGrew, 1992; Wrangham *et al.*, 1994; Rendell & Whitehead, 2001) have been willing to label what they do as ethnography, in the original sense of the term, as the descriptive and noninterpretative study of individual cultures (Winick, 1960, p. 193). In this sense, Goodall's (1968) account of the chimpanzees of Gombe is an ethnography, whereas her later *magnum opus* (Goodall, 1986b) tends toward ethnology, as it is interpretive and includes comparison of findings with other populations of wild chimpanzees.

Goodall's later volume also affords examples of the three levels of data available to cultural primatologists. *Nominal* data report presence or absence of a phenomenon, e.g. that Gombe's apes eat driver ants while Mahale's do not. *Ordinal* data allow qualitative comparison (i.e. more or less than). For example, Gombe's males do more charging displays than do females. *Interval* data can be assigned to a scale that affords arithmetical treatment, e.g. the percentage of time that dominant individuals groom others versus being groomed by them. Clearly interval-level data are preferred to ordinal to nominal on the ground of precision alone, regardless of one's preferences about positivism or reductionism. However, ethnology can be done on nominal data, as shown for humans in any issue of the journal *American Ethnologist*, or for nonhumans by Whiten *et al.*'s (1999) comparison of seven groups of wild chimpanzees.

An enduring problem of ethnography is the unit of study. This is reflected in the numbers from human studies: Price's (1990) 3500+ cultural groups versus Murdock and Provost's (1973) 186 groups. It is a matter of lumping versus splitting: too many nonindependent datapoints (Chiracahua versus Mescalero Apache: are they really that different?), and the sample size is inflated. Too few datapoints (are all Agta really the same, even if only some have women hunters?), and precious information is masked and lost. The former error of splitting will lead to 'false positive' (or Type-1) errors, in which a difference is claimed when none actually exists. This is because statistical tests are sensitive to sample size: 2/3, 20/30, and 200/300 are all the same percentage – 67 per cent – but only the latter yields a statistically significant difference. The latter error of lumping will lead to 'false negative' (or Type-2) errors, in which a real difference fails to be recognised.

For chimpanzees, there are three levels of analysis: subspecies, population, and community, etc. (see Chapter 1). As defined before, 'subspecies' is too gross ($n =$ only 3 or 4), but 'community' is too specific; thus, population is the most useful level for ethnographic description as a cultural unit. However, it is not always clear, in the absence of clear zoogeographical barriers to migration (gene flow) when one population ends and another begins. The unit groups (B, K, L, M, N) on the west side of the Mahale Mountains, may or may not send migrants to the groups to the east of the mountains, in Ugalla (Shimada, 2003). Lacking evidence, Mahale and Ugalla are treated separately here. Even within a small population, this distinction is not always clear. Goodall (2003) reports that the Gombe population of chimpanzees now numbers fewer than a hundred in three groups: north, central, and south. The well-studied central community (Kasakela) probably blocks access between the northern and southern communities. Similar issues arise in Taï (where there are five contiguous groups under study) but not in Bossou (only one group, isolated by 6 km from the next nearest group). For analytical purposes, only one population, Kibale, is here subdivided: into Kanyawara and Ngogo. This is done because the two are not neighbours and apparently do not exchange migrants.

In ethnography, the problem of change over time is solved by designating arbitrarily 1950 as the endpoint to the 'ethnographic present'. Thus, global influences that transformed traditional societies in the latter half of the twentieth century are set aside from comparative analysis. Thus, we can still 'study' gathering and hunting peoples, although no human group currently practises full-time this mode of subsistence. Were this not to be the case, and we used the real time of the twenty-first century, then drinking Cola-flavoured fizzy drinks might be seen as a human universal. By the same token, human cultural groups extinguished in the historical past (e.g. the Tasmanians) are still included in the ethnographic present, for comparative purposes.

Thus, using the same reasoning for chimpanzees, we can still make use of older ethnographic data on the apes' response to Cape buffalo at Gombe, although these large herbivores have since been shot out. Similarly, we can make use of archival behavioural data on the K-group at Mahale, though it was exterminated by its neighbours, M-group, in the 1970s.

Table 6.1 presents a complete list of wild populations of chimpanzees for which some data of potential cultural relevance have been published, and so are accessible as ethnography. For each population is given its subspecies, country, habituation status, and a published reference.

A population is labelled by the name of the study site associated with it by publication in scientific journals. Most have geographic referents and only a few have changed names (Kasoje is now Mahale, Impenetrable Forest is now Bwindi). Sites are divided into *long-term* (at least a year's continuous study) and *short-term* (less than that). In fact, there is a continuum that ranges from minutes (Beatty, 1951) to days (Kortlandt, 1962) to months (Tutin & Fernandez, 1985) to years (Yamagiwa *et al.*, 1996) to decades (Goodall, 2003). Beatty (1951) once saw one chimpanzee use a hammer and anvil to crack nuts; Goodall (2003) has followed four generations of chimpanzees over 44 years. Long-term sites are given in all capitals.

The traditional three subspecies are given here as central (*Pan t. troglodytes*), eastern (*P. t. schweinfurthii*) and western (*P. t. verus*). The only present candidate site from the supposed range of the central–western subspecies (P. t. *vellerosus*) is that of Gashaka, in eastern Nigeria (Sommer *et al.*, 2003). As has been the case since the beginning, the central African chimpanzees of the Congo basin rainforests remain under-represented, which is all the more disappointing since they are the most numerous and occupy the core range of the species.

Only seven of the populations are classed as *habituated* (i.e. tolerant of human observers at close range) so that focal-subject sampling of behaviour is possible. The two pioneering studies – at Gombe and Mahale – used provisioning at the outset (see discussion below) in order to accelerate habituation, but the other five did not. The long-term studies at Bossou and Taï were the first to show that habituation could be achieved without provisioning, and the latter-day studies of Kanyawara, Budongo, and Ngogo have built on these precedents. The remaining long-term sites range from some habituation (e.g. Goualougo), to seasonal progress (e.g. Assirik), to none, e.g. Lopé.

Fifteen countries across Africa – from Uganda in the east to Senegal in the west – have yielded ethnographic data on wild chimpanzees. The other countries yet to be heard from are troubled or have few, if any, chimpanzees left (Burundi, Cabinda (Angola), Ghana, Guinea-Bissau, Rwanda, Sudan). The two countries with the most populations studied are Tanzania – which also has the oldest study sites, at Gombe and Mahale – and Uganda, which has recently seen a burgeoning of new studies, plus one trail-blazing site, now rejuvenated, at Budongo.

Choosing one reference per site is not easy, especially from places such as Gombe, which has seen scores of publications. Listed in Table 6.1 are definitive references (Boesch & Boesch-Achermann, 2000), classic ones (Nissen, 1931), or ones that relate specifically to culture (Sept, 1992).

Table 6.2 lists populations of free-ranging chimpanzees, either confined to islands or to large enclosures or released into the wild (feral), usually after some rehabilitation (Hannah & McGrew, 1989). These are not wild chimpanzees, as most are constrained by barriers and artificial feeding, but they do forage, fight, and reproduce under African skies. Sometimes the control exercised by humans allows for naturalistic 'experiments', e.g. releasing a knowledgeable tool-user into a previously naïve population and watching the spread of the habit (Hannah & McGrew, 1987). Sometimes such populations alert us to phenomena that are later found in nature: a toolset was first seen

Table 6.1 *Sites in Africa where wild chimpanzees have been studied. Long-term sites in upper case. C = central;*
E = eastern; W = western.

Site	Subspecies	Country	Habituated?	Reference
ASSIRIK	W	Senegal		McGrew *et al.*, 1981
Ayamiken	C	Equatorial Guinea		Jones & Sabater Pi, 1971
Bafing	W	Mali		Moore, 1985
Banco	W	Ivory Coast		Hladik & Viroben, 1974
Belinga	C	Gabon		Tutin & Fernandez, 1985
Beni	E	Congo (DRC)		Kortlandt, 1962
BOSSOU	W	Guinea	+	Sugiyama & Koman, 1979a
BUDONGO	E	Uganda	+	Reynolds, 1992
BWINDI	E	Uganda		Stanford, 2002
'Cameroon'	C	Cameroon		Merfield & Miller, 1956
Campo	C	Cameroon		Sugiyama, 1985
Cape Palmas	W	Liberia/Ivory Coast		Savage & Wyman, 1844
Chambura	E	Uganda		Wrangham, pers. comm.
Diecke	W	Guinea		Matsuzawa *et al.*, 1999
Dipikar	C	Equatorial Guinea		Jones & Sabater Pi, 1971
Filabanga	E	Tanzania		Itani & Suzuki, 1967
FONGOLI	W	Senegal		Pruetz, 2001
GASHAKA	C-W	Nigeria		Sommer *et al.*, 2003
GOMBE	E	Tanzania	+	Goodall, 1986b
GOUALOUGO	C	Congo (RC)		Morgan & Sanz, 2003
Ishasha	E	Congo (DRC)		Sept, 1992
Kabogo	E	Tanzania		Azuma & Toyoshima, 1961–2

Site		Country		Reference
KAHUZI-BIEGA	E	Congo (DRC)		Yamagiwa et al., 1996
KALINZU	E	Uganda		Hashimoto et al., 2001
Kanka Sili	W	Guinea		Kortlandt, 1986
Kanton	W	Liberia		Kortlandt & Holzhaus, 1987
KANYAWARA (KIBALE)	E	Uganda	+	Wrangham et al., 1996
Kasakati	E	Tanzania		Suzuki, 1969
Kilimi	W	Sierra Leone		Harding, 1984
Kindia	W	Guinea		Nissen, 1931
'Liberia'	W	Liberia		Beatty, 1951
LOPE	C	Gabon		Tutin et al., 1991
MAHALE	E	Tanzania	+	Nishida, 1990
Mbomo	C	Congo (RC)		Fay & Carroll, 1994
Ndakan	C	Central African Rep.		Fay & Carroll, 1994
NDOKI	C	Congo (RC)		Kuroda et al., 1996
NGOGO (KIBALE)	E	Uganda	+	Watts & Mitani, 2002a
Ngoubunga	C	Central African Republic		Fay & Carroll, 1994
Nimba	W	Guinea–Ivory Coast		Matsuzawa & Yamakoshi, 1996
Odzala	C	Congo (RC)		Devos et al., 2002
Okorobiko	C	Equatorial Guinea		Jones & Sabater Pi, 1971
Sapo	W	Liberia		Anderson et al., 1983
SEMLIKI	E	Uganda		Hunt & McGrew, 2002
TAÏ	W	Ivory Coast	+	Boesch & Boesch-Achermann, 2000
Tenkere	W	Sierra Leone		Alp, 1997
Tiwai	W	Sierra Leone		Whitesides, 1985
Tongo	E	Congo (DRC)		Lanjouw, 2002
Ugalla	E	Tanzania		Itani, 1979
West Cameroon	C	Cameroon		Struhsaker & Hunkeler, 1971

Table 6.2 *Sites in Africa where captive or released chimpanzees have been studied.*

Site	Subspecies	Country	References
Abuko	*Ptv*	Gambia	Brewer, 1978
Assirik	*Ptv*	Senegal	Brewer, 1982
Baboon Islands	*Ptv*	Gambia	Brewer & McGrew, 1990
Bassa	*Ptv*	Liberia	Hannah & McGrew, 1987
Conkouati	*Ptt*	Congo (RC)	Tutin *et al.*, 2001
Ipassa	*Ptt*	Gabon	Hladik, 1973
Rubondo	*Pt*	Tanzania	Borner, 1985
Tacugama	*Ptt*	Sierra Leone	Alp, 1999

on the Baboon Islands in the Gambia (Brewer & McGrew, 1990) but has since been seen in nature (Sugiyama, 1997).

If Tables 6.1 and 6.2 tell us what we have to work with as ethnographers on a nonhuman species, what does it amount to? With about fifty sites, in which at least some data relevant to culture have been collected, we have for the chimpanzee probably the largest number of cultural groups available for comparison of any species of nonhumans. This is almost double the number (twenty-seven) listed just over a decade ago (McGrew, 1992, pp. 178–9), indicating the continuing growth of field studies of wild chimpanzees. (For other primates, it seems likely that the number of field sites of study of the superspecies of savanna baboons or of the Japanese macaque would rank second.)

Further, although the geographical species range of the chimpanzee across Africa pales by comparison with, for example, the superspecies of vervet monkeys, the range of ecological variation encompassed by the apes is greater. Chimpanzees live in a range of habitats from rainforest (rainfall of more than 3000 mm annually) to savanna (800 mm), (McGrew, 1992, p. 124). Although the chimpanzee's latitudinal range essentially is equatorial – being less than 20° from north to south – the species' longitudinal range is almost 50° from west to east – from Senegal to Uganda. Altitudinal variation is also great, with chimpanzees living from sea-level in Gabon to at least 3300 m elevation in the Congo (DRC) (Yamagiwa *et al.*, 1996). Finally, because they cannot swim, chimpanzee distribution is broken up by water, especially by lakes in east Africa, and rivers flowing into the Atlantic in central and west Africa. All of these ecological factors provide a fertile foundation for cultural variation.

In terms of time, the accumulated years of field study of chimpanzees is impressive. Whiten *et al.*'s (1999) comparison of six field sites drew from a total of 151 years of research. With the addition of two more sites – Assirik and Lope – Whiten *et al.* (2001) pushed the total to more than 170 years of study. At least five field sites (Bossou, Gombe, Kibale, Mahale, and Taï) have been in *continuous* operation for more than 20 years. Only the field sites studying Japanese macaques (e.g. Koshima or Arashiyama) exceed such scientific longevity (Matsuzawa, 2003). It may be no coincidence that this species of monkey is so well-known for its cultural heritage. The chimpanzee can be contrasted with the bonobo: no population of the latter ever has been habituated fully to daylong observation, and no bonobo study site has managed long spells of continuous research, given the periodic political unrest in the DRC (Kinshasa). It seems likely that bonobo culture eventually will prove to be equivalent to chimpanzee culture, if time and circumstances permit the ethnography to be done (Hohmann & Fruth, 2003).

PROVISIONING

When sociocultural anthropologists do field work among traditional societies, the problem of the visitors affecting, even distorting, the lives of the residents is ever-present. Even the proverbial fly on the wall may exert an influence on nearby humans, so the effect must be much greater when one human being sets out to study another. In a society in which a hollow ostrich egg is the primary container of water, an empty tin can discarded by visitors may become a revolutionary vessel. There are many parallels in the initial contacts of ethnographers with their subjects, whether these be fellow humans or apes.

A vexing example is provisioning, i.e. researchers providing prized objects, often foodstuffs, to curry favour with the subjects of study (Asquith, 1989; Williamson & Feistner, 2003). For chimpanzees, this is usually calorie-rich vegetation, e.g. banana, sugarcane, or citrus. The advantages can be immediate and almost magical: when Goodall (1971) offered bananas to the Gombe chimpanzees, they almost swamped her camp with their enthusiasm. So much fruit was distributed that sympatric baboons became accidentally habituated, too. At Wamba, the volume of sugar cane is apparent from the forest floor being littered with spat-out fibre, as is seen in almost all published photographs of bonobos from the site (de Waal & Lanting, 1997). Close-range behavioural details that might have taken years to achieve without provisioning were obtained instead in months. This benefit is not to be sneezed at in a world in which research funding for long-term projects is hard to get. Habituation is a prerequisite for high-profile, wider, recognition, whether in *National Geographic* magazine or on the Discovery channel.

But the costs of provisioning are great too, both scientifically and ethically (Williamson & Feistner, 2003). Consuming a day's worth of energy in a few minutes of gobbling down calorific cultigens is bound to affect activity budgeting, especially if foraging and ranging widely to do so are rendered superfluous. This may have unexpected consequences: at Gombe, heavily provisional chimpanzees used their new leisure time to prey on baboons, also attracted to the bananas. Before and after such provisioning, baboons barely figured in the apes' diet (Goodall, 1986b). Thus, the chimpanzees showed cultural adaptation to incorporate a new prey species that had been there all along, but was now made vulnerable in the open feeding area.

More pointedly, concentration of prized food in a limited space (the provisioning site) led to increased aggression among Gombe's chimpanzees (Wrangham, 1974). Furthermore, the irresistible bananas seemed to draw chimpanzees from far and wide, thus increasing social density. This is not likely to be confined to Gombe: at Wamba, the superabundance of sugar-cane may have led to the high frequencies of food–sex orgies of bonobos. At Mahale, chimpanzees were called in to partake of provisioned food, by humans summoning them for that purpose, thus entraining their movements by instrumental conditioning. All of these aspects of social life seem likely to have influenced the fission–fusion society of the apes, although there are few real data to test this.

Ethical considerations vary for the above effects of provisioning, but reservations are unavoidable when they entail matters of life and death. This may happen indirectly or directly. Indirectly, wild primates may be further encouraged to raid crops if provisioned with them elsewhere. This may get them injured or killed as pests. Primates who come to tolerate humans and their material culture at close range may be more likely to be taken by hunters or trapped in snares. Directly, the single most hazardous effect of provisioning may be transmission of pathogens, whether microbes or parasites (Wallis & Lee, 1999). In nature, it is hard to establish cause and effect when epidemics break out, but the death rate of Gombe's chimpanzees from infectious diseases appears to be high, ranging from poliomyelitis to measles to mange.

Happily, provisioning has all but died out in field primatology, and so its contaminating effects on observational data in the ethnographic record of primates is less and less of an issue. Even more happily, some of the purported negative side-effects of provisioning turn out to be nonproblems. Power (1991) blamed provisioning for incidents of fatal aggression in wild chimpanzees, but such aggression, whether within or across groups, turns out to be equally common in chimpanzee groups that have never been provisioned (Kerbis Peterhans et al., 1993; Wrangham, 1999; Muller, 2002).

Perhaps, in a methodologically ideal world, the ethnography of wild chimpanzees would be done with no interference by human observers. Technically, this could now be done with telemeterised implants, minicams, and satellite uplinks. The apes would go about their daily affairs under surveillance, never aware that they were being watched, and over time a complete picture of their lives would come together. But just as no ethnographer of human beings would find this way of working satisfactory, so, I suspect, would any cultural primatologist. If offered the trade-off of observer absence and cleaner data, versus messier data and the observer present, most primatologists would probably opt for the latter.

TRADITION

Tradition, in any meaningful sense of the word, entails a collective body of knowledge and action that shows continuity over time within a group. (This usage differs from that of making tradition merely equivalent to social learning in animal behaviour, as discussed in Chapter 4.) But how much time is long enough? Traditional continuity regenerated again and again over time, as opposed to genetic or environmental stasis, is maintained by transmission of culture from older to younger individuals. Such vertical transmission may be from parent to offspring, as is expressed, sometimes idealistically (and simplistically), for example, in phrases like 'mother tongue' or 'imbibed with mother's milk'. But culture also can pass from grandparent to grandchild, older to younger sibling, aunt to niece, etc., as well as between nonrelatives.

This suggests that genuine tradition reflects continuity over generations, rather than over any arbitrary period of months or years. Generation time reflects species-typical life history, especially the period from birth to reproductive maturity, after which a new generation can be produced. For primates, generation time varies from less than 2 years for marmosets to more than a decade for chimpanzees. This makes cultural primatology much easier to study with some primates than with others, especially if longitudinal data are needed. If we are interested in innovation, we can only recognise it with confidence against a background of its known absence. For example, when Georgia began to do the grooming hand-clasp at Yerkes, it was known to be new, because the behavioural history of all the chimpanzees in her group up to that point was known (de Waal & Seres, 1997). Thus, to appreciate the emergence and endurance of a tradition, we must collect ontogenetic ethnographic data.

Of course, nontraditional culture is not subject to these constraints. Peer-to-peer transmission of culture, horizontally, can be nearly instantaneous, as the global phenomenon of hip hop music makes clear. This can make ethnography difficult, when culture is so transient. By the time you read this, Britney Spears may be just as outmoded as Mott the Hoople.

There is little or no evidence of horizontal cultural transmission by chimpanzees in nature, although Matsuzawa's (1994) study of Yo at Bossou comes close. Yo was a female immigrant, probably from nearby Nimba, who brought knowledge of cracking *Coula* nuts, a foodstuff unknown in the Bossou chimpanzees' diet. Transmission was shown by providing artificially *Coula* nuts at the outdoor laboratory. Resident apes ignored the uninteresting objects, but Yo set to work cracking them, and this food-processing habit was acquired by younger residents (Matsuzawa, 2003).

Thus, the existence of real tradition is not a necessary condition for ethnography, but it is a sufficient one. But this refinement dodges the major question of *content*: tradition of what? Ethnography of what?

DOING APE ETHNOGRAPHY

Tylor's (1871) definition of culture included knowledge, belief, art, law, morals, custom, and any other capabilities and habits. This resembles Ingold's (2001) list of 130 years later: intentions, purposes, values, orientations, perception, memory, and contexts of action. The two have in common at least two elements: one trivial, and one profound. The trivial common element is their inclusiveness. Culture permeates everything that a culture-bearer does, from acting to thinking to feeling. The profound common element is the extent to which culture can be studied directly, i.e. what proportion of the above is observable and what proportion must be inferred. Most must be inferred: Tylor's beliefs and Ingold's purposes are mental states that never can be verified.

The implication of the inclusiveness of culture recalls the old saw: 'Something that explains everything, explains nothing.' To say that everything a person does is cultural, and that culture explains everything a person does, is gloriously circular. This is where the scientific method comes in, to look for causal relations between independent and dependent variables. This is why reductionism is necessary to explain a phenomenon, rather than just to muse about it.

The seemingly contradictory implication of the immateriality of much (most? all?) of culture is that standard methods of scientific enquiry will never be enough. We can record, count, measure, dissect, etc., Tylor's (behavioural) habits, but what about knowledge? We can do the same for Ingold's (physical) contexts, but what about memories? This challenge applies equally to humans and nonhumans, for reasons given in Chapter 1. Humans are not off the ethnographic hook just because they can be interviewed before, during, or after, they act. Chimpanzees are not ruled out of ethnographic court, just because they (apparently) do not verbalise about their actions.

All that we can do with both species is to draw the strongest possible inference in any given case. We can lay out hypothesised premises specifically and explicitly and attend to their consequences. As iterations accumulate, we can refine supposed relationships between variables, starting with correlation and moving toward causation. Such methods are no less applicable to verbal behaviour than to nonverbal, whether the latter be vocalisation, facial expression, gesture, or posture. They are equally applicable to evanescent acts (e.g. playing a flute) as to enduring objects of action, e.g. the flute itself.

Consider the following imaginary example. In a particular African forest, the human inhabitants do not appear to eat a species of duiker. These people eat other ungulates, and the duiker species is eaten by other predators. It is an available and palatable prey, yet it is ignored, even when it accidentally ends up in a hunter's snare. A sociocultural anthropologist asks the people why they decline to eat this type of antelope, and they say it is forbidden to do so. They elaborate that the duiker is a clan totem and consumption of it would lead to sterility. No matter how coherent and compelling is their story, it may not explain the dietary omission or reflect reality. The people may eat the duiker meat in secret. Or they may eat the meat overtly, but relabel it as something else. Or they may eat the meat in special circumstances that do not hold at the time of the anthropologist's interrogation. And so on.

Or, more seriously, they may not eat the duiker for simple ecological reasons, e.g. because there are other mammals in the forest that provide more meat for less effort or at less risk or in a shorter time (Alvard, 2003). That is, whether people articulate it or not, or are even aware of it or not, optimal foraging theory may explain why they do not eat this species of duiker. This does not negate the folklore account obtained from discourse between informant and ethnographer, but that account is not an explanation. Happily, most of the time, -emic and -etic (Harris, 1979) accounts of people's actions are congruent, or at least can be reconciled, but when they are not, we are left to make inferences based on the best data at hand. (See Aunger, 1994, for a biosocial treatment of food taboos.)

Now consider a corresponding hypothetical example for chimpanzees. In the same forest, the apes also do not eat this species of duiker. They eat other species of ungulate, as do the people. The chimpanzees even catch young duikers occasionally in the course of their daily foraging, but instead of eating them, they play with them! Cultural primatologists cannot interview the subjects so, instead, they must test alternative hypotheses based on ethological and ecological data. It may turn out that the same optimality explanation that worked for the people also applies to the apes. But suppose that none of the environmental hypotheses apply, what then? It may be that social or cultural variables will account for the omission from the apes' diet. These are testable too. Suppose that every time a chimpanzee youngster starts to sniff or mouth one of the duiker playthings, her mother gives her a soft threat bark, or an older sibling takes the duiker away from the youngster. Suppose, when a party encounters a scavengeable duiker carcass, perhaps cached by a leopard, all the adults give it a wide berth, avoiding even looking at the potential foodstuff. And so on.

So, which of the two hypothetical situations is a taboo, and therefore cultural, and so amenable to ethnography? Is the human example any less cultural because it makes ecological sense? Is the nonhuman example any less cultural because it lacks a verbal report? In fact, all other aspects (anthropomorphism, speciesism, anthropocentrism, chimpocentrism (see McGrew, 1992, p. 216)) aside, we are left with similar data calling for a similar conditional inference. More data on the same or different variables will clarify the situation, and raise or lower the probabilities of correct inference, but a taboo always will be inferred, no matter what the species.

These arguments about the prospects and pitfalls of doing ethnography with the closest living relations of humankind apply across the board. They apply to aspects of life as environmentally grounded as seeking food, drink, and shelter, and to aspects as socially embedded as competing for status, finding a mate, and raising offspring. They apply to elementary technology that will last for aeons and to momentary social nuances that signal an individual's social identity. They apply to habits that are chimpanzee universals (and therefore are indicative of 'chimpanzee nature') and chimpanzee uniquenesses. The next two chapters are a survey of these phenomena, the results of ethnography, with some telling examples chosen for in-depth treatment.

7   Chimpanzee material culture

Happily for the cultural primatologist, chimpanzees make use of material culture every day of their lives. Such elementary technology is by far the most accessible manifestation of culture in these apes. Furthermore, material culture by definition leaves a tangible record of its use as an enabler, facilitator, or product of behaviour. However humble or mundane are the objects involved, the scientific advantages of concreteness cannot be overestimated.

Elementary technology is the knowledgeable use of one or more physical objects as a means to achieve an end. I distinguish between technology as elementary when *simple*, and advanced when *complex*, using the terms in Oswalt's (1976) sense. This is different from the object being an end in itself. The grapefruit-size fruit of the woody *Saba* vine may be opened systematically and skilfully (Corp & Byrne, 2002), so that its juicy pulp is eaten, but this is food-processing, not technology. If the same fruit were to be plucked and thrown at a companion, then this use would be technology, for the fruit becomes a missile aimed to strike another. Or, if after consumption of its contents, the empty shell of the fruit were to be used by the ape as a container to collect dripping rainwater, then the item would be transformed into an artefact.

If such technology meets the definitional criteria posed in Chapter 2, then it is material culture. That is, if it is standardized in a collective way that is characteristic of the group, then it is cultural. This can be contrasted with the mindless techniques of some other species of object users, e.g. the stereotyped and instinctive use of pebbles by parasitoid wasps to tamp down the earth at the entrance to their burrows (Tinbergen, 1951). It can also be contrasted with the nonsocially learned tool use of other animals, e.g. the prising tools of

Galapagos woodpecker finches (Tebbich *et al.*, 2001). Such creatures have efficient and essential techniques, but not technology (Ingold, 1986).

In his treatise on material culture, Oswalt (1976) distinguished between artefacts and naturefacts. A naturefact is a natural form, used without prior modification, while an artefact is the end-product of modification of an object to fulfil a useful purpose. Either can be cultural or noncultural, and distinguishing between them is not always easy. A boulder used as an anvil upon which to crack hard-shelled fruit is a naturefact; a flexible probe of vegetation, cut to size and stripped of leaves, used to fish-out social insects, is an artefact. But what of the boulder that becomes pitted with wear after untold repetitions of being struck by fruits? It is modified, but only inadvertently, and so remains a naturefact. Also, what about the perfect stem, of just the right natural length, width and configuration, that can be used as a fishing-probe without alteration? It must remain a naturefact, strictly speaking, for it was used as found. This may seem pedantic, but in certain cases (e.g. the problematic brush-stick (see below)), such distinctions are crucial.

Oswalt (1976, p. 199) also elucidated four principles for the production of artefacts: reduction, conjunction, replication, and linkage. Table 7.1 elaborates on these, and gives human and ape examples (see also McGrew, 1992, pp. 131–44).

The importance of raw materials needs emphasising. Chimpanzee material culture relies on natural materials, either from living creatures or nonorganic resources. There are no synthetic artefacts, unless wild apes have been 'corrupted' by contact with humans, e.g. chimpanzees at Gombe chew and suck ('wadge') cardboard cartons. The most famous example was of a challenging adult male, called Mike, who banged together discarded empty paraffin tins as part of his agonistic displays (Goodall, 1971). This portable drumming greatly enhanced the effectiveness of the display, and he rose to the top of the male hierarchy, but once the tins were removed, the habit died out, as there was no natural source of these objects.

Table 7.1 *Principles of production of artefacts (Oswalt, 1976, p. 199).*

		Examples	
Principle	Definition	Human	Chimpanzee
Reduction	Reduce mass of form to make functioning product	Flaked stone	Stripped fishing probe
Conjunction	Combine different parts to create finished form	Hafted axe	Lined nest
Replication	Craft similar structured units to function as one form	Prongs of leister	Compressed leaf sponge
Linkage	Use physically distinct forms in functional combination	Bow and arrow	Dipping wand and bent-over sapling

Most chimpanzee material culture is made of plant matter: twig, bark, vine, leaf, stem, shoot, root, bough. Much less is made of animal matter: bone, skin. Of nonorganic matter, only water and stone figure. This predominance by vegetation means that most of chimpanzee culture is perishable, especially under the warm and humid conditions of the tropics. Most of the archaeological record of apes is lost within their lifetimes, and much of it disappears within days or weeks. The notable exception is lithic technology – mostly hammers and anvils – which is amenable to much more timedepth in analysis (Joulian, 1994; Mercader et al., 2002).

Before plunging into material culture in chimpanzees, we should recall its costs and benefits to the researcher: the main benefit is that material culture, when recognised, can be studied even with totally unhabituated subjects, i.e. in the absence of its users (McGrew et al., 2003). This is almost the only way for most populations of wild apes to be included in cultural primatology, and so it is a godsend. (The only other way is through direct analysis of food remnants, which can be used to reconstruct their processing. This potential avenue of analysis seems untapped, but ethological data suggest that it could be done (Byrne & Byrne, 1993; Corp & Byrne, 2002).) Furthermore, persisting concrete objects can be measured, archived and re-analysed at leisure; they can be assembled and compared, and if properly curated, this can be done years, decades, or millennia later. This contrasts sharply with momentary behavioural events, which if not recorded, are lost for ever. Actions, in their evanescence, can never be verified or re-examined, except when recorded by tape or film.

The main cost of material culture is that an object alone is silent as to its use. We can never know, for example, how it was gripped, or even if it was used only by hand. The oral tool-use of orangutans could not have been predicted by the tools themselves (O'Malley & McGrew, 2000; van Schaik & Knott, 2001). This is exactly the problem that palaeoanthropologists face in trying to know the function of, for example, Acheulean bifaces ('hand-axe'). They have never seen them used and, short of a time machine, never will, so can only infer their use based on archaeological research that mimics hypothesised functions, e.g. butchering carcasses (Schick & Toth, 1993). Cultural primatologists at least can work toward habituation and eventual observation of the acts that accompany the objects.

A related cost to relying on objects is that they may be missed if the field worker has never seen that type of tool used before, elsewhere. This is a good reason for an aspiring chimpologist to start a career by working with one of the few fully habituated populations of these apes. (I was lucky to start field work in 1972–5 at Gombe and Mahale, before moving on to unhabituated subjects in other parts of Africa.) Failure to see technological objects may be especially problematic with naturefacts: an anvil used to smash baobab fruits may reveal itself only from minute amounts of the green fuzz that comes from the outer layer of the fruit. This residue can be mistaken easily for lichen, and so go unrecognised (see Bermejo *et al.*, 1989, for an apparent example with termite-fishing probes).

This chapter starts with universals of chimpanzee material culture, but moves on to types that vary across populations by presence or absence, to types that show nuanced differences cross-culturally, to a few types that are unique to only one population of apes. These will be discussed in terms of their varied functions in daily life, from subsistence to social, to self-maintenance. Finally, the importance of material culture to chimpanzees will be assessed along with its limitations. Much of this chapter should be seen as an update to my earlier book, *Chimpanzee Material Culture* (McGrew, 1992).

SHELTER

All great apes in nature are born in a shelter and those that succumb to illness or injury (as opposed to predation or pongicide) may die there (Yamagiwa, 1998). In between, each individual beyond the age of weaning makes at least one shelter a day, and most make more, every day of their lives, for decades. This is the most solid of norms. It would be just as bizarre for a chimpanzee to sleep on a bare branch as it would be for a human to do so, in contrast to the lesser apes, monkeys, and prosimians. Prosimians do not make shelters, but do improve existing cavities (e.g. by lining tree-holes with leaves) as do other mammals. This is a clear distinction across the primate order, between the large-brained hominoids and the others.

These constructed shelters of great apes have been labelled variously sleeping platform, bed, nest. The former excludes terrestrial shelters but only the latter is misleading, as it suggests a residence for central place foraging or rearing young or storing food, as is common in many vertebrates (Hansell, 1984, 2004). None of these happens in a chimpanzee shelter, which is rarely used for more than one bout of resting, whether by day or night. Unfortunately, however, the use of the term 'nest' has become the norm in primatology, at least in English, since the first usage in ethnography by Nissen (1931). I follow that precedent here, reluctantly.

The making of a nest looks simple, and it is certainly quick: in 3–5 min. on average an ape bends or breaks branches inwards to form a roughly circular frame about 1 m in diameter. These main branches are interwoven and their side branches tucked in, and the interior of this shallow dish is lined with detached leafy twigs. The resulting shelter may be slept in for about 12 h from dusk to dawn, or during the day for naps of a few hours. There is nothing more predictable in chimpanzee daily life than this universal behavioural sequence and its artefactual outcome. It is the cornerstone of chimpanzee nature.

It is therefore surprising how little actually is known about nests and nest-building. The best description of the making of the chimpanzee nest is still Goodall's (1962) first published article. Forty years later, no one has bettered it (see Fruth & Hohmann, 1994, 1996, for the more recent standard-setting studies on the subject). In captivity, where the nature and nurture of nest-building is best tackled, there is nothing in print since Bernstein's (1967, 1969) studies almost as long ago as Goodall's. There is no real cross-cultural comparison: Whiten *et al.* (1999, 2001) ignored nest-building in chimpanzees, on the grounds that it showed no cross-populational variation. (This uniformity seems unlikely, especially as van Schaik *et al.*, 2003, found several types of such variation in wild orangutans.)

Most astonishingly, there has been no study of the *function* of the chimpanzee nest. It is a shelter, but from what? Hypothesised functions are antipredation, thermoregulation, antipathogen, and mental health. The antipredation hypothesis states that arboreal sleeping is more secure than terrestrial, being better protection against nocturnal carnivores. But an animal can be arboreal without being a shelter-maker: sympatric adult male baboons perch on limbs, while smaller-bodied juvenile apes build nests. The thermoregulation hypothesis states that a pallet provides insulation against overnight hypothermia; but if so, why build nests on warm and dry nights? (At Assirik, the overnight low temperature averaged more than 25 °C in some months, but the chimpanzees carried on making nests (McGrew *et al.*, 1981).) The antipathogen hypothesis states that high up in the trees, disease vectors (e.g. mosquitoes) will be less common. But if so, what about gorillas who typically sleep on the ground? Finally, the mental health or cognitive hypothesis says that with their big brains, apes, like humans, need to dream. Dreaming requires rapid eye movement (REM) sleep, with its overall muscle relaxation, and so a bough-perching chimpanzee who lapsed into dreaming might fall out of a tree! Hence, the nest as a crib. None of these hypotheses has been tested, but all are testable, some even with captive subjects. A good starting-point would be actualistic nest-building and overnight occupation by a keen human researcher. If that person also slept on the ground as a control, and (1) survived, (2) recorded skin temperature by sleeping naked with thermocouples, (3) counted mosquito bites, and (4) kept a dream diary or wore a telemeterised (electroencephalograph) (EEG) rig, a good start would be made to learning about the function of nest-building.

Lots of differences exist across populations of wild chimpanzees, but it is not clear if any of these are cultural. For example, the height of the nest above ground varies from only a few metres at savanna sites to a few tens of metres at rainforest sites. At first glance, this difference seems impressive, until one looks at the heights of trees in both types of habitat, which are correspondingly low and high. (For an early example of such a comparison, see Baldwin *et al.*, 1981.) A similar argument can be applied to variation in frequency of compound nests, i.e. nests that combine foliage from more than one tree. This can be seen to reflect density of individual trees and degree of closure of the canopy. Such ecological constraints are obviously outside the control of the apes, yet do influence their shelters.

Crude measures of the nests themselves (e.g. dimensions) show no obvious differences that cannot be explained by raw materials, e.g. the species of woody plants available. What is needed is an architectural study that deconstructs nests into their component boughs, branches, and twigs in relation to one another. Computer simulation might aid in this. When certain woody species seem to be preferred as nest sites, then the structural characteristics of the vegetation (e.g. tensile strength) need to be made clear. Chimpanzees may choose to build nests more often in *Parkia* trees because they have thick, soft leaflets, or because they are sited in breezy spots, or merely because it is a common species.

Against all expectations, some populations often nest on the ground, e.g. at Bwindi (Maughan & Stanford, 2001) and Nimba (Humle & Matsuzawa, 2001). There is no absence of suitable trees, so are these populations relieved of predator pressure? Perhaps they are scared of heights? Only further study will say.

It may be that chimpanzee nests are stereotyped in construction and composition like those of some other animals (Hansell, 1984). It may be that a nest is a nest is a nest, regardless of which individual or group or subspecies is studied. It may be that nest-building techniques develop in an individual's lifetime through some combination of neuromuscular maturation and trial-and-error learning. No one has yet looked at the nests of youngsters to see if they resemble those of their mothers or other close companions. No one has yet looked to see if individuals who often nest together in overnight sleeping parties also make more similar nests. No one has yet looked at the nests of orphans, to see – if lacking a regular maternal model – their nests are more atypical.

Yet there is evidence of social influences on nest-making, and even of social learning. In the wild, youngsters up to the age of weaning (5–6 years old) typically share their mother's nest at night, and so never build a night nest of their own. This changes, literally, overnight when a younger sibling is born. From the very next night onwards, the now displaced older offspring makes her own nest, independently. Self-responsibility for shelter for the youngster is achieved suddenly and smoothly, with no time for trial and error. Of course, infants practice at building day-nests before this, but day-nests are flimsy and insufficient for overnight, or so they seem. No one has compared them systematically yet.

From captivity comes evidence of the need for social learning, as nursery-reared chimpanzees seem unable to make proper nests

(Bernstein, 1967, 1969). Although there are signs of innate tendencies (e.g. pulling in objects to sit on them, or placing objects about them at arm's length ('magic circles')) this never leads to real nest-building (Ladygina-Kohts, 2002). On the other hand, wild-reared chimpanzees in captivity will seek to build nests, even when substandard, unnatural raw materials are all that is available, e.g. cut branches, hay, burlap bags, etc. Once these wild-born individuals die, and so are no longer there to pass on nest-building to the next generation, it seems likely that the cornerstone of chimpanzee nature will disappear in captive populations.

SUBSISTENCE: FAUNIVORY

The first anecdotal records of behavioural patterns as candidates for material culture had to do with subsistence (Savage & Wyman, 1844; Beatty, 1951; Merfield & Miller, 1956). Soon after Goodall's (1962) pioneering findings on nests came her discoveries on faunivory (Goodall, 1963). Since then hundreds of publications have appeared on the material culture of chimpanzee subsistence, here defined as use of objects to find, get, process, and eat food, or drink. Some of the best-known and most telling patterns of food-getting are as follows.

(1) *Termite-fishing* (see Figures 7.1, 7.2 and 7.3) is the use of a flexible probe of vegetation to extract termites from their terrestrial, earthen, mounds. In east Africa, Goodall (1964, 1968) stressed the motor skills needed to thread the tool into the hole, then to winkle out the underground insects that have attacked the intruding object, affixing themselves to it with their mandibles. As in angling, the prey are sensed by touch. The task is low-energy, self-paced, and almost risk-free, so it is ideal for mothers with offspring, and females predominate in this performance (McGrew, 1979). The payoff in calories and amino acids is notable (McGrew, 2001b). At the other end of the chimpanzee's species range – in far west Africa – the technique is similar but the tools are different: while Gombe's apes make most of their probes from grass and bark, Assirik's apes use

FIGURE 7.1 Chimpanzees at Gombe fish for termites. From left to right: Opening a hole with the forefinger; extracting insects with a grass blade probe; eating dabbed termites off the back of the hand. Photograph by Linda Marchant.

FIGURE 7.2 Adult male chimpanzee arrives at a mound at Gombe with tool already made for termite-fishing.

FIGURE 7.3 Two adult female chimpanzees fish for termites at Gombe while their offspring play beside the mound.

mostly twigs and leaf midribs (McGrew, 1992, p. 170). In rainforests in central Africa, chimpanzees fish for termites in combination with digging or perforating tools (McGrew *et al.*, 1979). Although large-bodied, fungus-growing, mound-building termites (Macrotermitinae) apparently are sympatric with chimpanzees throughout their range, some populations of the apes do not harvest them, e.g. in Taï (Whiten *et al.*, 1999).

(2) *Ant-fishing* is the use of the same sort of tool, but to extract instead wood-boring ants (*Camponotus* spp.) from their arboreal nests within living trees. Similar motor skills are needed, and often the fishing ape must support herself while suspended in the canopy. The pattern was described first at Mahale (Nishida, 1973; Nishida & Hiraiwa, 1982), but also occurs in central and western Africa (Whiten, *et al.*, 2001). Unlike termites, *Camponotus* ants are mostly indigestible

chitin, but their taste is strong, leading Nishida to suggest that they are eaten as tasty junk food. In termite-fishing, the seated apes have both hands free to use and usually show exclusive hand preference, left or right, to fish. In ant-fishing, the apes switch hands, as one limb becomes fatigued from supporting the weight of the body while the other hand fishes. *Camponotus* is a cosmopolitan genus, and probably is sympatric with chimpanzees everywhere in their range, so failure in some populations to capitalise on this ubiquitous prey is notable and suggests that they have yet to acquire the taste.

(3) *Ant-dipping* is the use of straight, stiff vegetation to collect massed driver ants. These are predatory, fiercely biting army ants that move from one bivouac to another, vacuuming the forest floor of animal prey. One type of ant-dipping entails using a long wand of stem or shoot to extract ants from their temporary nests (McGrew, 1974). It is a two-handed task, in which one hand holds the end of the tool in a power grip, while the other sweeps the length of the tool in a loose precision grip, accumulating the ants in a jumbled mass of several hundred. The other type of ant-dipping involves a short tool held in one hand and dipped directly into a foraging or migrating column on the move (Sugiyama *et al.*, 1988; Sugiyama, 1995b; Boesch & Boesch, 1990). These simple dips yield a few ants each time, nibbled directly by the lips from the tool. The two-handed technique was first described at Gombe and the one-handed at Taï, and it was thought to be an arbitrary cultural difference between sites (Whiten *et al.*, 2001). In some cases, only the artefacts inform us of driver-ant eating (Anderson *et al.*, 1983; Alp, 1993), and we are left to infer technique from tool.

More recently, Humle and Matsuzawa (2002) have shown this dichotomy to be simplistic: Bossou chimpanzees use both techniques, depending on which type of driver ant they are exploiting. The fierce ants are taken by the two-handed wand technique when concentrated at the nest; the less pugnacious form is taken by the one-handed sequence when on the move. It seems that we must re-examine which driver ants are present at which sites of chimpanzee study and which techniques are used to exploit them. Instead of cultural variation determining the harvesting habits of the apes, it may be that the prey (ants) are driving the behaviour of the predator (apes). On the other hand, Mitumba's chimpanzees at Gombe use two contrasting techniques on the same ants: they use the standard two-handed technique with the wand, as described for the Kasakela community (McGrew, 1974), but also grab handfuls of the ants, which they carry up into the trees to eat (Wallauer, personal communication).

Besides termites and ants, tools also are used sometimes by chimpanzees to harvest bees, either to extract the insects or their stores of pollen or honey from the nest. The latter is nature's purest energy source, and is just as prized by apes as it is by humans (McGrew, 2001b). Social insects have in common a high collective biomass, making their exploitation worth while for a large-bodied predator. Equally, they are daunting prey, because of their formidable physical defences (mound, elevated hive) or their painful antipredator tactics (bite, venomous sting). Elementary technology either makes accessible prey that are otherwise invulnerable (the termites' 'castles of clay') or makes more manageable or bearable the bites and stings of the prey.

Some ants are processed without tools, but using protocols that are likely to be cultural: *Crematogaster* ants are cryptic, living in the hollow dead stems of certain plants. Getting the ants by breaking off the dead plant parts and splitting them lengthwise is easy, once the predator knows that the ants are there. Mahale's chimpanzees are experts at this (Nishida, 1973). Weaver ants are the opposite: their presence is obvious from their pale orange colour and quick reactivity to disturbance. They swarm over intruders in defence of their shelters of living leaves bound together by larval silk. Chimpanzees rush in, pluck a leaf-bundle and retreat to safety. There quickly they crush the nest and its occupants, by rolling the natural container between the palms, or palm and sole. Then the leaves are picked away, one by one, and discarded, as the squashed ants are licked off. All that remains of the feast are a few leaves on the forest floor, trailing wisps of silk.

Chimpanzees also eat vertebrates, which they often get by hunting. They seek, detect, stalk, pursue, capture, and kill other mammals, especially monkeys and ungulates. Systematic comparison can be done across populations only on the hunting of red colobus monkeys. Everywhere that they co-occur with chimpanzees, red colobus are the primary prey, sometimes to devastating extents (Stanford, 1998a; Watts & Mitani, 2002a). Chimpanzees hunting monkeys may be solitary or social, but in all cases it is nontechnological, i.e. there are no weapons. The only use of tools in meat-eating is the customary use of twigs to pick out bone marrow and the occasional use of leaves to swab out brain fragments from skulls (Whiten *et al.*, 2001).

Striking differences do exist across populations in hunting practice: Gombe chimpanzees often hunt alone (Busse, 1978), while Taï and Mahale chimpanzees hunt socially (Boesch, 2002). Social hunting is thought to be co-operative, but to varying extents: *similarity* in hunting means doing the same acts; *synchrony* in hunting means temporal conjunction; *co-ordination* in hunting means congruence in both time and space; *collaboration* in hunting entails different but complementary roles. In collaborative hunting at Taï, an individual may act as driver, blocker, chaser, or ambusher, and all hunters share in the kill. At Mahale, there is no collaboration, and alpha males control the kill and the distribution of the meat is constrained (Nishida *et al.*, 1992; Marchant, 2002). Finally, at Gombe, the Kasakela chimpanzees specialise in preying on young monkeys – often an infant snatched from the mother's belly – while at Taï, adult monkeys are the primary prey. Taï is exceptional in this way, as other sites resemble Gombe in focusing on immatures (Mahale: Nishida *et al.*, 1983; Ngogo: Mitani & Watts, 1999).

Some of these interpopulational differences reflect environmental differences across sites: In high, closed-canopy evergreen forests such as Taï, the monkeys have multiple escape routes in all directions. This makes it hard for one or a few chimpanzee hunters to 'corner' a prey. In lower, broken-canopy woodland, as at Gombe, monkeys may be trapped in isolated trees and so can be subdued, even by a lone hunter. Thus, collaborative hunting pays at Taï, but seems maintainable only by relatively egalitarian distribution of meat (Boesch, 2002). Differences in choice of prey by age are less obviously environmentally driven, but the demography of the colobus is an obvious factor to check (Stanford, 1998a).

SUBSISTENCE: HERBIVORY

Chimpanzees are omnivores, and most of their food is plant matter, not animal matter. Of the various vegetative and reproductive plant parts, fruit is the most important, at least for energy. But fruits vary in edibility; few are eaten whole. Most are 'processed' by spitting out or defecating seeds, fibre, skin, rind, stems, etc., while consuming the 'pulp' (usually mesocarp). Only Corp and Byrne (2002) have done a detailed study of manual fruit-processing, and there are likely to be many as-yet-unstudied cultural differences in fruit-handling.

Fruit smashing is percussive technology: Large, hard-shelled fruits are smashed against a hard substrate, or anvil. This is done with fruits large enough to grasp but too large to crush in the mouth. Goodall (1968) first described fruit-smashing for *Strychnos* spp., a wide-ranging genus with orange-sized spherical fruits, by the apes of Gombe, but little since has been published on it (McGrew *et al.*, 1999).

More unusual is the smashing of the ovoid football-sized fruits of the baobab by arid country chimpanzees in far west Africa (McGrew *et al.*, 2003; Marchant & McGrew, 2004). Chimpanzee and baobab are sympatric only at a few savanna field sites, but at Assirik in Senegal, baobab is a keystone foodstuff (Baldwin 1979). Baobab seeds are among the most nutritious of wild-plant foods eaten by human gatherer-hunters: they are high in both crude protein (36 per cent) and fat (29 per cent) (Murray *et al.*, 2001). The fruit's matrix also offers useful amounts of simple sugars (39 per cent), calcium and vitamin C. On the other hand, the seeds must be processed to exclude the seed coat, which has high levels of trypsin inhibitor.

In Senegal, at both Assirik and Fongoli, chimpanzees pluck the baobab fruit by detaching the long stem (*c.*50 cm) at the proximal end. Then they hold that end in one hand and swing the fruit over-head and then down in an overarm motion that maximises cen-tripetal force. The fruit is struck against a bough (arboreal) or a root or stone (terrestrial) and the impact sound is audible for several hundred metres. (Brewer, 1982, provides an illustration of the technique.) Once cracked, the shell is split open, revealing the white matrix, embed-ded in fibre, and containing the seeds. When immature, the matrix is chewy and the seeds are pale green and soft; the seeds are split open in the mouth and the kernels eaten and coats spat out. When mature, the matrix is chalky and tastes of ascorbic acid; it is sold for human con-sumption in local markets. The seeds are then brown and hard. Assirik chimpanzees swallow them whole, then retrieve the partly digested seeds from their faeces for reingestion. This is not coprophagy per se, but it illustrates how much the apes value this food item. Hadza gatherer-hunters in Tanzania retrieve baobab seeds from baboon fae-ces, wash and dry them, then pulverize them with stone hammer and anvil before eating them (Murray et al., 2001).

Baobabs afford another opportunity for elementary technology to be used to get the fruits. The tree's trunk is broad and cylindrical with smooth bark and no primate can climb a large baobab unassisted. At Assirik, chimpanzees and baboons can get access to the baobab fruit only by climbing into the crown of a neighbouring tree that overlaps with the baobab's crown. Human harvesters use poles with footholds as ladders. In captivity, chimpanzees spontaneously invent and use pole ladders (Menzel, 1972, 1973a) and in nature, rehabilitated chim-panzees can be taught easily to use them to get baobabs (Brewer, 1978), but the only case of a wild chimpanzee using a ladder was at Fongoli, in Senegal, where a chimpanzee made use of a ladder left behind in a baobab tree by a human honey-collector (unpublished data).

While fruits are smashed on anvils, nuts are cracked with ham-mer and anvil. The hammers may be stone or wood (branches), and anvils may be stone or living roots. Although nut-cracking is known in several far west African rainforests (Sierra Leone: Whitesides, 1985; Liberia: Anderson et al., 1983; Guinea: Sugiyama & Koman, 1979b;

Ivory Coast: Joulian, 1996), two sites – Bossou and Taï – provide most of the behavioural data.

At Bossou, one community of chimpanzees regularly cracks the nuts of the oil palm, using hammers and anvils of stone (Sugiyama & Koman, 1979b), but neighbouring groups also are known to crack palm nuts (Humle & Matsuzawa, 2004). There are no other species of nut-bearing trees available in their small range of primary forest, so that, for example, they have no natural opportunity to crack *Coula edulis* or *Panda oleosa* nuts. Their nut-cracking is well-studied in an 'outdoor laboratory', a cleared area in the chimpanzees' home range, where nuts and raw materials for elementary technology are made available (Sakura & Matsuzawa, 1991). The experimental introduction of *Coula* nuts revealed how a single innovator or new local user – Yo – could provide the source of new cultural transmission (Matsuzawa, 1994). The ideal observational conditions enable, for example, the hand preference of the Bossou chimpanzees in nut-cracking to be determined: all are completely lateralised, i.e. each uses only the left or right hand to hold the hammer (Sugiyama *et al.*, 1993).

At Taï, chimpanzees use hammers of stone or wood to crack various species of nuts on anvils of stone or root (Boesch, 1978). *Coula* nuts are common, relatively soft-shelled, and can be cracked with wood; *Panda* nuts are scarce, hard-shelled, and can be cracked only with stone. There is a correlation between the hardness of the nuts and the appropriate raw material for cracking them (Boesch & Boesch, 1983). Furthermore, quartzite, which makes the finest hammer stones, is rare in the lowland rainforest, so stones are carried from worksite to worksite in strategic ways (Boesch & Boesch, 1984a). Nut-cracking is a skilled motor activity but in a different way from termite-fishing: appropriate force must be applied precisely to the nut if it is not to pulverize, resist the blow, or carom away (Boesch & Boesch, 1993). It is an energetically profitable, seasonal, activity (Guenther & Boesch, 1993), done with manual precision, like termite-fishing. Also like termite-fishing, it is self-paced and low-risk, making it offspring-friendly, so it is not surprising to see female predominance

(Boesch & Boesch, 1984b). As at Bossou, nut-cracking is lateralised manually at Taï (Boesch, 1991a).

One population – Bossou – uses percussive technology in another, striking, way: *pestle-pounding* (Yamakoshi & Sugiyama, 1995). Here, the prey is again the oil palm, but it is the apical growth point ('heart') that is processed and eaten. The ape detaches a young frond, after having isolated it in the crown; this takes great strength. The soft pith is eaten, but then the frond becomes a pestle to be pounded into the mortar of the palm's heart. This produces a rich, stringy soup of pulp that is eaten by hand. Contrary to expectations (McGrew, 2003b), this does not kill the palm, and so is a sustainable form of harvesting (personal observation).

Chimpanzees also use elementary technology to get drinking water (e.g. Sugiyama, 1995a). Goodall (1968) described Gombe chimpanzees using green leaves crushed together as a sponge to get water from tree-holes. Leaf-sponging is now known to be a chimpanzee-universal (Whiten *et al.*, 2001). Bossou chimpanzees combine eating and drinking when they take green algae from pools using stick tools (Matsuzawa, 2003). Algae scooping is unique to Bossou, perhaps because such pools are found rarely elsewhere (Whiten *et al.*, 2001). The most enigmatic form of drinking technology is well-digging. Chimpanzees in savanna habitats in Senegal and Uganda dig wells in sandy stream-beds (Hunt & McGrew, 2002). Interestingly, they do so even when there is some surface water present, but only when it is not flowing. This suggests that they are avoiding the potential health hazards of infectious disease by not drinking stagnant water. Whether chimpanzees use digging sticks or not to make the wells remains to be seen, but it is known that chimpanzees can dig up succulent tubers by hand without tools if the soil is suitably loose (Lanjouw, 2002).

SOCIAL MATERIAL CULTURE

Chimpanzees use objects as weapons, either as missiles or as clubs (Goodall, 1964). Stones and broken-off branches are thrown at other apes, humans, competitors (baboon), predators (leopard), or prey (bush pig). In addition to this *aimed throwing* – seen customarily at most field sites – there is *unaimed throwing* as part of the male charging display. Any object in the displayer's path is likely to be picked up, broken off, or uprooted, and flung about (Goodall, 1968).

Use of clubs is less common and is not seen customarily at any site, but some of the Kanyawara males club females in a savage and prolonged way (Linden, 2002). If this occurred in humans, we would call it spousal abuse.

While it is not weapon use, *branch-dragging* occurs in the noisy charging display that is a male chimpanzee-universal, being found in all eight populations for which data are available (Whiten *et al.*, 2001). Branch-dragging is also the commonest form of object use in bonobos (Hohmann & Fruth, 2003).

Objects are used in courtship, mostly by males, in a variety of ways, all of which seem to function to get the female's attention. *Shaking branches* in courtship is a chimpanzee-universal, being habitual or customary everywhere that has been tested in systematic ethnology (Whiten *et al.*, 2001). At Mahale, amorous males clip dead leaves in the lips, making a characteristic tearing sound (Nishida, 1980b), but elsewhere in Africa, *leaf-clipping* may signal tension or play (Whiten *et al.*, 2001). Pulling a stem noisily through the clenched hand, so that the leaves are stripped away noisily, serves a similar function, as does knocking with the knuckles on a root or branch, or slapping a branch.

Leaf-grooming is an enigmatic behavioural pattern that has no obvious direct function. Why would a chimpanzee spend time and effort cleaning an inanimate object, complete with accompanying lip-smacking and close-up scrutiny? As set out in McGrew (1992, p. 188) its hypothesised indirect functions are attention-getting, grooming-stimulating, or doodling, but none of these has been tested properly yet. It is hard to think of an explanation that would *not* be cultural, given its arbitrariness and patchy distribution, i.e. it is present at all five east African sites but absent at all four non-east African ones (Whiten *et al.*, 2001).

Chimpanzees also incorporate objects into play, in ways that suggest cultural content, e.g. social games. One of the seven universal patterns that emerged in Whiten *et al.*'s (2001) analysis was *play start*. In this pattern, a play *provocateur* holds leaves or twigs or some other object in the mouth and flaunts these to the prospective play partner (Goodall, 1986b). This leads to chasing play, and sometimes to role reversal, if the pursuer snatches the object and then flees with it. In one community, the material culture of play includes a pattern that may be both social and nonsocial play: *leaf-pile pulling* (Nishida & Wallauer, 2003). In the dry season, youngsters in M-group at Mahale pile up leaves and then pull them along the bare ground, often moving backwards in front of others. The authors liken it to a solo dance, being a performance displayed to others. Leaves are available everywhere, yet the pattern is seen elsewhere only at Gombe, where it is habitual, not customary.

Finally, the most ubiquitous of all sixty-four candidate patterns in Whiten *et al.* (2001) was *drumming*. It is the only such pattern to be customary in all nine populations, sampled from Senegal to Uganda. Surprisingly, little systematic analysis has been done on drumming across populations (cf. Arcadi *et al.*, 2003). We do not know if any information content is contained in a series of beats, or if there are dialectic differences across communities, populations, or subspecies. Two idiosyncratic findings give us cause for optimism: Boesch (1991b) found at Taï that an alpha male – Brutus – was using drumming to signal information about time and space. His drum-beats told others of direction of movement in the forest and of duration of resting sessions, and this enabled them to find him. Sadly, when Brutus died, the habit died-out too. Arcadi and Mugurusi (2003) analysed both the drumming and the ground-slapping of a percussive adult male at Kanyawara. The two types of acoustic communication showed many similarities in form but also some significant differences in performance. In both cases, we need more data to know if these patterns go beyond idiosyncrasy to more generalised cultural significance.

Finally, there is the *'raindance'*, a sometimes spectacular collective display, done just before a storm (in response to thunder and lightning?) or during rainfall (in frustration at being drenched?) Saplings or lianas are bent and flailed or thrashed, and sometimes swapped off, leaving a trail of damaged vegetation. (For a graphic portrayal, see Goodall, 1967, pp. 82–3). Interestingly, with one exception (Taï), the raindance is an east African custom, being done at Budongo, Gombe, Kibale, and Mahale, but not further west in Africa. Thus, material culture appears in all aspects of chimpanzee social life: communication, aggression, courtship, grooming, and play.

SELF-MAINTENANCE

The humblest form of day-to-day cultural activities are those that make an individual's life safer, more comfortable, healthier, more efficient, or happier. These solutions to life's little problems may arise from individual tinkering by trial-and-error. Such idiosyncrasies can be a function of morphology, temperament, or intelligence, and need not have anything to do with others. But here I focus only on those patterns of behaviour that meet cultural criteria. The primary one – nest-building – has been covered already.

The most striking use of material culture in self-maintenance is *leaf-swallowing*, a form of oral self-medication (Huffman, 1997; Huffman & Hirata, 2003). It was first described by Wrangham and Nishida (1983) for *Aspilia* leaves at Gombe and Mahale, based on two types of data: atypical ingestive behaviour, and faecal samples. In the former, leaves were consumed singly and deliberately and swallowed without chewing. In the latter, leaves emerged undamaged and folded, sometimes with intestinal parasitic worms caught on the Velcro-like surface of the leaves. Also, leaf-swallowing is often done first thing in the morning, on an empty stomach. The phenomenon is found at thirteen sites across Africa, from Uganda to Senegal, but is best-known from Budongo, Gombe, Kibale, and Mahale.

There are differences across regions and local similarities in the plant species chosen for swallowing. These suggest cultural dissemination and diffusion. Most investigations stop with reporting the field data, but Huffman and Hirata (2003) introduced suitable leaves to naïve captive chimpanzees in Japan to see if they would use them. The apes re-invented the custom and showed apparent observational learning.

If this is self-medication, how do we know it works? Obviously, short of a double-blind, placebo-controlled, random-assignment-to-treatment trial, we cannot. What we can do is look at symptom relief: if an individual with signs of parasitic infection seeks out a particular species of plant with known antihelminth properties, and if those symptoms decline or disappear after ingestion, then we have a prima facie case (Huffman & Seifu, 1989).

Also, how can we imagine that a self-medicating ape knows what it is doing? It seems improbable that an ape would have conscious knowledge of the cause and effect of such treatment. Happily, that is not needed. If ingesting a particular leaf in a particular way at a particular time leads to perceptible relief, then operant conditioning can shape an adaptive behavioural pattern without consciousness or intentionally being required.

Investigatory probing with a tool of vegetation that is sniffed after making contact with an unknown or strange object is a chimpanzee-universal. Such probes may be used to check cavities or to touch a dead body. Also a universal is *branch-clasp grooming*, in which mutual groomers each drape an upper limb over an overhead branch, while grooming the other with the other hand. Such a postural configuration is apparently the basis for the *grooming hand-clasp* (see Chapter 8). Although frequencies vary for these patterns across populations (Whiten *et al.*, 2001), no one has yet done systematic comparisons to look for cultural nuances. This reflects the tendency for cultural primatologists to pay more attention to dramatic, rather than to subtle, variation.

Chimpanzees use vegetation in hygiene, with the most common pattern being leaves used as *napkins* to wipe off bodily fluids, or plant juices. The former may be blood, semen, faeces, urine, or mucosal secretions; the latter may be sap, sticky seeds or pulp

(Goodall, 1968). At Kibale, the apes use leaves to *dab* at wounds, in inspection.

Closely related to this is the chimpanzee's treatment of ectoparasites like lice or ticks. At Gombe, chimpanzees place a groomed-off parasite on a leaf, and use that as a platform upon which to squash it (Boesch, 1996). At Budongo, the parasite is placed on a leaf for inspection, but is not squashed (Assersohn *et al.*, 2003). At Taï, the parasite is placed on the groomer's forearm, not on any object, and then squashed (Whiten *et al.*, 2001). At other sites, the offending invertebrate simply is swallowed, trusting to digestive juices to do the job. Finally – for airborne insect pests – chimpanzees at Taï and Budongo habitually use a leafy twig as a fly-whisk.

Chimpanzees also use material culture to make life more comfortable. Big leaves piled on the wet ground make sitting on the substrate more comfortable while doing terrestrial tasks (Hirata *et al.*, 1998). When harvesting kapok leaves for food, there is the problem of big thorns on the tree's trunk and boughs. Tenkere's chimpanzees solve this by using leafy twigs as 'sandals' to protect their feet and as 'cushions' to allow them to sit and eat (Alp, 1997). Comfort is in the eye of the beholder: an adult male at Mahale sticks twigs up his nose to trigger sneezes, then eats the mucous (Nishida & Nakamura, 1993). Perhaps not surprisingly, few group members have followed his lead (Marchant & McGrew, 1999).

Finally, there is one group of chimpanzees – the Kasakela community at Gombe – in which objects are used habitually in playful self-stimulation. Infants tickle themselves with stones or sticks (Goodall, 1986b). It has never been seen elsewhere.

Thus, material culture is part of every aspect of chimpanzee life: subsistence, sociality, and self-maintenance. Is there any other species, except humans, that uses elementary technology in so many aspects of life?

SIGNIFICANCE OF MATERIAL CULTURE

Chimpanzees have toolkits. They are generalists at material culture, not one-shot wonders, as with so many other specialist animal species.

Their use of elementary technology varies from the simply functional (e.g. nut-cracking) to the possibly symbolic (e.g. leaf-clipping) and many points in between. The size of the toolkit of any particular population of apes seems to be a function of length of study and degree of observability. Artefacts should be obvious, regardless of degree of habituation of the apes, but even artefacts can be missed if the field researcher lacks the 'search image' for the modest tools made of natural vegetation.

Chimpanzees also use toolsets (i.e. a sequence of tools used to achieve a single goal) in which the right order of use, e.g. A – B – C – D, is essential for success (Brewer & McGrew, 1990). A simple human example is processing a coconut: first you (A), drain the juice by puncturing the 'eyes', then you (B) break it open with a hammer, then you (C) scrape out the meat with a blade. There is no point in doing A last, and you cannot do C before B. Sugiyama (1997) has relabelled tool *sets* as tool *composites*, but this confusingly combines two different types because there is a difference between using two or more tools in succession versus simultaneously. For toolsets, a chimpanzee seeking honey may use a chisel, bodkin, and dipstick (Brewer & McGrew, 1990), which is comparable to coconut-processing. Similarly, a chimpanzee may sit on a bent-over sapling and use it as a vantage point for ant-dipping, just as a human can use a ladder for painting a ceiling. Only the latter should be called a 'tool composite', following Oswalt's (1976) definition of linkage, as in Table 7.1.

Is all chimpanzee material culture of the same depth and breadth? It would be surprising, as the material cultures of human traditional societies vary so greatly. The foraging Tasmanians lacked hafted stone tools, bone tools, nets, fish hooks, shields, spear throwers, boomerangs, canoes, dogs, and fire-making. Yet all of these were present in the cultures of their aboriginal counterparts on the Australian mainland (McGrew, 1987).

A well-known pattern of variation in material culture in chimpanzees is that of hammer and anvil use across Africa. Ten of the sites in Table 6.1 (pp. 92–3) have nut-cracking: Banco, Taï, Cape Palmas in

Ivory Coast; Bossou and Nimba in Guinea; Kanton, 'Liberia', Sapo in Liberia, Tiwai in Sierra Leone. All are of *Pan troglodytes verus* in west Africa and only one is east of the Sassandra-Nzo river in southwestern Ivory Coast. That site is Banco, a small forest block in the capital city of Abidjan; it may be that the nut-crackers there are released pets (Joulian, 1996). Furthermore, eight of the ten sites are not far from a focal point near the convergence of the borders of eastern Liberia, western Ivory Coast, and southeastern Guinea. This suggests that most of the nut-cracking results from diffusion from a single invention. Also, there is ecological survey (as opposed to behavioural or artefactual) evidence too, from both Ivory Coast (Boesch *et al.*, 1994) and Guinea (Matsuzawa *et al.*, 1999), and this gives the same picture. Boesch *et al.* (1994) surveyed systematically and found no nut-cracking sites to the east of the Sassandra-Nzo river.

What are we to make of the absence of nut-cracking by the chimpanzee of central and eastern Africa? At Lopé, in central Africa, all of the ingredients are there: multiple nut species and plenty of stones, sticks and roots for hammers and anvils (McGrew *et al.*, 1997). Yet there is no evidence of nut-cracking by the apes at Lopé, despite all the elements being available.

A sceptic might observe that the Lopé chimpanzees are unhabituated, and so the evidence of absence might actually be absence of evidence (or, a false-negative error). This objection will not apply to the chimpanzees of Gombe, in east Africa, where after 40+ years of focal-subject sampling and nest-to-nest data collection, not a single sign of nut-cracking by the apes has been seen (McGrew, 1992). This holds, despite the fact that oil palm mesocarp is the single most common item in the Gombe chimpanzees' diet (Wrangham, 1977), and that local people use local stones to crack the nuts in the same forest (McGrew, 1992, p. 207). Similarly, nut-cracking notably is absent at the second-oldest site: Mahale. It is hard to explain these regional differences between west and non-west Africa on environmental grounds but, of course, since the habitats differ across sites, an ecological explanation can never be ruled out. For example, the non-western

chimpanzees may derive their energy and nutrients from eating termites, a habit that is absent from all the nut-cracking sites. Perhaps
on optimal foraging grounds, termites are to be preferred to nuts.

However, further scrutiny of the east African material culture seems to falsify that hypothesis: at none of the seven Ugandan
sites (Budongo, Bwindi, Kalinzu, Kanyawara, Ngogo, Semliki, and
Chambura) in Table 6.1 are *either* termites *or* nuts harvested. Termites
are present at all Ugandan sites, and crackable nuts apparently are
present at some, yet there is almost no extractive foraging, for animal
or plant prey (Whiten *et al.*, 2001). It is not that Ugandan chimpanzees
are nontechnological (see Table 9.1, p. 179); their social and self-
maintenance tool use matches that of their Tanzanian cousins. What
is missing is extractive technology in subsistence, and in that sense
their food-getting is primitive by comparison to their peers in central
and western Africa. In subsistence terms, they are the equivalents
to the Tasmanians, but just as the Tasmanians had rich social lives
comparable to those of other antipodean foragers, so do the Ugandan
chimpanzees.

What are the limits of chimpanzee material culture? Some basic
items of even the simplest human material culture are absent, e.g. the
container. Chimpanzees have access to the raw materials that humans
use to make containers (e.g. bark, vine, skin) but do not do so. But
what would a nomadic, eat-as-you-go quadruped do with a container
anyway? Without the freed hands of a biped, there is no way to carry a
container. Without a home base, there is nowhere to cache a container.
Without food that requires off-site processing before consumption,
there is no need to use a container to collect it. Interestingly, captive
chimpanzees placed in settings in which containers are useful (e.g. to
carry liquids) learn to use them readily and spontaneously (McGrew,
1992, p. 221). Wild chimpanzees no more invent unnecessary tools
than Efe pygmies in tropical rainforests invent ice-picks.

It is sometimes said that a major difference between humans
and apes is that only humans use tools to make other tools (metatechnology). Thus, a knapped stone blade may be used to shave an

arrow shaft. At first, this may seem an unbridgeable gap, but it seems less a difference in kind and more a difference in degree. First, chimpanzee dentition often acts as an equivalent technology to human lithics, whether to clip with incisors, puncture and slice with canines, or crush with molars. This is how chimpanzees 'butcher' monkeys, in ways that evolutionarily reduced human dentition no longer allows. Further, these same robust teeth also process many raw materials of plant matter, whether to make wands for ant-dipping or probes for termite-fishing.

Second, chimpanzees sometimes do use tools to modify other tools, if by 'modify' one means reorient or reposition. A leaf sponge lost out of reach in a tree hole may be retrieved with a stick, so that sponging may continue (unpublished data). One of the hypothesised means by which brush-sticks are made is by crushing the end of a tool between hammer and anvil (Sugiyama, 1985). However, this has yet to be seen and brush-stick-making sites have yet to be found. Alternative hypotheses are that the tools are made by crushing the sticks with the molars, or that the brushy end is just fraying produced by repeated insertions into the hole during its use.

However, the only habitual pattern of metatechnology is the anvil-propping found at Bossou (Whiten et al., 2001). In this case, an anvil stone is repositioned by propping it up with another stone, resulting in its working surface becoming more horizontal (Matsuzawa, 1991). This makes the anvil more efficient, as a nut resting upon it is more likely to stay put when struck with the hammerstone. On less than horizontal surfaces, nuts roll off the anvil or carom off its tilting surface when hit.

Another generalisation about elementary technology found commonly in textbooks (e.g. Relethford, 2000) is that only humans depend on material culture, while material culture in other species is some sort of optional bonus. The logical way to test the hypothesis is to look for a human society that lacks material culture. If found, it falsifies the assertion. Sure enough, such a search reveals no exceptions: all human societies have material culture, suggesting dependency.

But the same test for chimpanzee societies reveals the same result: no chimpanzee population has yet been found that lacks elementary technology. (The apparent exceptions are short-term studies of only days or weeks in duration). Thus, unless we find a chimpanzee group without material culture, it seems that the apes are just as dependent upon it as are humans.

Finally, lest it be thought that material culture is some holy grail, to be sought until achieved, or is some panacea, to be always a source of only benefits, it should be said that material culture is a mixed blessing. There are plenty of examples of human technology having both positive and negative outcomes, from the domestication of fire onwards. The price we pay for the warmth of a wood-burning stove is a polluted atmosphere. The modern satellite-driven technology of deep-sea fishing may be raping the oceans and extinguishing species, but we can eat our fish sticks whenever we want.

There may be a chimpanzee equivalent. At Gombe, the Kasakela community is killing red colobus monkeys at an unsustainable rate (Stanford, 1998a). On one occasion, the apes killed a quarter of a troop's membership in *one hunt*. On an annual basis, the colobus troops at the core of the chimpanzee community's range are being hammered so hard that local extinction is a possibility. (It seems likely that the apes have already exterminated the duikers at Gombe.) However, this over-exploitation is not unique to Gombe; it seems to be occurring at Ngogo, too.

The monkey slaughter puts paid to another textbook generalisation: that only humans have culture to the extent of modifying the resources in their environment. Such effects via material culture are concrete and obvious. But what about nonmaterial culture (e.g. the hyperpredation on red colobus monkeys) which leaves no artefacts, except in the patterning of neurons firing in the brain? The next chapter tackles evanescent culture in chimpanzees.

8   Chimpanzee society

Recently, there has been a spate of reports on 'eavesdropping' in animals, in which individuals act differently, depending on who is watching or listening, and later modify their actions based on what they have seen or heard (Whitfield, 2002). For example, Siamese fighting fish pay close attention when their neighbours fight, and tailor their later interactions with winners and losers accordingly. Furthermore, defeated males prefer to court females who did not witness their humiliation. These impressive social achievements in such humble creatures yield several implications.

One is that, yet again, nonprimates and nonmammals show abilities that will surprise most primatologists, who in their blinkered existence often look for comparison only to humans, and not to fellow vertebrates. We are forced to acknowledge that one can only appreciate where human and nonhuman primates stand in relation to one another by taking a wider view.

A second implication is that nonhuman society is about more than behaviour, interaction, relationship, and social structure (Hinde, 1987). It is also about individuals calibrating their actions in relation to vicarious knowledge of others' behaviour, interactions, and relationships. Thus, by watching B, A learns not just how B acts, but also how B fits-in relative to B's interactants and relations. This is the basis for real society based on social cognition.

The most important implication is that eavesdropping creatures have the potential for culture based on collectivity. In acquiring knowledge from the acts of one's fellows, there is an emergent fellowship based not just on common interest (e.g. co-operative hunting), but also on mutual fate, e.g. sharing the prey according to co-operation.

In principle, this means that cultural primatologists are not confined to material culture. When they have unrestricted access to the full repertoire of chimpanzee behaviour on an individual basis in good observational conditions, they need only to be able to record accurately. Making a permanent record on film or tape is even better, but it is not necessary. This is no different from a sociocultural anthropologist listening to the verbal output of an informant or filming a funeral rite.

In practice, there is little ethnology done on nonmaterial culture in chimpanzees. In Whiten *et al.*'s (2001) list of thirty-nine behavioural patterns that show putative cultural variation, only one – hand-clasp – is neither a subsistence nor a self-maintenance activity. Moreover, Whiten *et al.* (1999, 2001) declined to consider what is arguably the richest source of nonmaterial culture: vocalisation. Studying acoustic communication requires specialist knowledge and skills that most primatologists lack. Any chimpologist can tell a pant-grunt from a pant-hoot, but few know their way around a sound spectrograph.

VOCAL COMMUNICATION

The first serious study of vocal communication in chimpanzees was probably that of Marler (1969) at Gombe. He recorded, labelled and classified as many types of calls as possible, and the quality of the data was good enough to distinguish between individuals (Marler & Hobbett, 1975). Because no other study site produced comparable records, the idea that chimpanzee calls were stereotyped and invariant remained intact until a student of Marler's, John Mitani, took up the problem. Mitani started with Asian apes, especially the singing gibbons, but in visiting several chimpanzee field sites, starting with Mahale, he encountered interpopulational variation. Mitani *et al.* (1992) found that elements of the chimpanzee's loudest call – the pant-hoot used for long-distance communication – varied across populations. Such variation, when found in songbirds, was termed 'dialectical', following linguistic precedents for such variation, and so the term was also applied to apes.

Once opened in the 1990s, the floodgates of vocal communication research on chimpanzees yielded a torrent. Mitani and Gros-Louis (1995) found sex differences in screams in both chimpanzee and bonobo. The selective factors responsible for individuality in vocalisations were elucidated (Mitani *et al.*, 1996), as were those aspects of the environment that shape long-distance transmission of vocal information (Mitani & Stuht, 1998). At the same time, social effects were demonstrated both in nature and in captivity: chorusing causes convergence, so group, as well individual, differences have their origins in social influences (Mitani & Gros-Louis, 1998; Marshall *et al.*, 1999).

More recently, there has been a retrenchment in explaining geographic variation in chimpanzee vocalisation, at least for long calls: social factors have been joined by nonsocial ones, some even as basic as the psychophysics of sound attenuation by vegetation (Mitani *et al.*, 1999). Because no two habitats are alike, any variation between populations in their vocalising could be explained by the habitat. There is an ongoing debate, as yet unsettled (e.g. Arcadi, 1996, 2000) as to the relative contributions made to vocal output by social learning and by environmental constraints. This could in principle be settled by translocation experiments, in which apes of various ages are exchanged across various environments, from closed rainforests to open savannas. However, these are probably not feasible and would be ethically objectionable to many. The upshot of vocal communication in chimpanzees is that it is influenced socially in ways that, if we saw similar variation in other areas of chimpanzee life, we would term it 'cultural'. On the other hand, we are reminded that no behavioural pattern exhibited by any organism developes in an environmental vacuum.

GROOMING

Social grooming occurs in almost all primates, all terrestrial mammals, and in many vertebrates. This is not surprising, for two reasons, one crucial and the other practical. Crucially, hygiene relates to health, and health relates to reproductive success, both directly (lactation) and indirectly (mate-guarding). Since grooming enhances hygiene, especially maintenance of the protective body surface, it makes sense. But why not groom oneself, instead of having to depend on others? The answer is simple: in every organism there are parts of the body's surface that cannot be seen or felt, or which cannot be scrutinised or manipulated finely by oneself. No primate can see the lice eggs at the base of hairs on the back of her own neck. So, we depend on others.

Social grooming may be one-sided: the mother who cleans carefully her infant, from tip to toe, does not receive grooming from her offspring. Hygiene rules supreme here, as parental investment; healthy offspring are more likely to thrive. More interesting is reciprocal grooming, in which A grooms B, and then later (seconds, minutes, hours, or days) B grooms A. (Most social grooming is one-to-one, but not all.) Such turn-taking is characteristic of monkeys and apes, where it has been well documented (see, for example, Dunbar, 1992, 1996). Reciprocal grooming allows for social accounting (debt, interest, risky investment) in the primary currency (grooming itself) or in other services that can be exchanged at a rate. (How many minutes of grooming equals a leg of freshly killed monkey?) Grooming is likely to be the basic currency, since it is universal and everyday, but some ledgers may last a lifetime, and some may not balance (Dunbar & Sharman, 1984). Thus, politics may explain social grooming between adults, whether this relates to sex or power, or both.

Only chimpanzees and bonobos – but not gorillas (Williamson, personal communication) – engage in the most exotic form of social grooming, which is mutual. That is, A grooms B at the same time that B grooms A. Why is this simple simultaneous exchange so rare? Perhaps because it is more demanding cognitively, as it is basically multitasking, occupying two roles (groomer and groomee) at once. (This is not true roleplay in the strict sense; however, see below.) Perhaps more telling is the implication that in mutual grooming, the individual is committed to a partnership. The dyad is the simplest social grouping, and this sociality suggests potential for culture. When we groom reciprocally, we are individuals interacting: you present your shoulder, then I groom it. When we groom mutually, our every act is complementary: I cannot groom your head if you are grooming my back. Thus, mutual grooming is interdependent. Moreover, unlike other primates, chimpanzees engage in social grooming cliques and clusters, which is far more complex than the usual grooming dyad (Nakamura, 2003).

So, if social grooming, especially mutual grooming, is the stuff of chimpanzee social life, then we might expect it to be based on social learning, and therefore amenable to cultural processes. There is some evidence of social learning, but of the crudest kind: chimpanzees reared artificially in species-isolation (i.e. deprived socially) do not spontaneously show social grooming and fail to learn it later. However, they are also intellectually and emotionally damaged from also being sensorially, physically, nutritionally and otherwise deprived. Thus, they may fail to exhibit normal behavioural patterns for any number of reasons.

Another problem with seeking cultural processes in social grooming is that it is so commonplace. Like nest-building (see above), its universality is perceived as uniformity. It is hard to believe that no detailed study of variation in the form of social grooming has been done for chimpanzees. One example will serve to show its potential: chimpanzees are well known to accompany social grooming with specific noises, e.g. lip-smack, tongue-click, teeth-clack, the 'Bronx cheer', or raspberry, etc. (Marchant, personal communication). These are so arbitrary as to be highly likely to be socially learned, yet these vocalisations have only begun to be studied (Nishida *et al.*, 2004).

SOCIAL SCRATCH

Consider the following behavioural pattern, so elementary as to be often ignored: scratching. All vertebrates scratch an itch, i.e. rub an appendage against their body surfaces to relieve a superficial irritation (Diezinger & Anderson, 1986). We humans do so reflexively and absent-mindedly, even when we are asleep, and so do chimpanzees, who rake the flexed fingertips repeatedly over the skin, often in long, rhythmic strokes. Self-scratching is a species-universal, seen in every chimpanzee in every population, captive or wild.

Only one population of chimpanzees takes it further and engages in social scratching: M-group at Mahale uses social scratch in social grooming on a daily basis (Nakamura *et al.*, 2000). The motor pattern is conspicuous and drawn out, almost in slow motion, as the groomer scratches the groomee's trunk or limbs, but especially the back. Not surprisingly, socially dominant individuals receive more back-scratching than do subordinates. The exception to this is when mothers socially scratch their young offspring; their respective social ranks are irrelevant.

Consider the implications of this simple interaction: unlike grooming-off a foreign object, there is no product to social scratch. There is no reward to the scratcher, as the itch (if there is one) belongs to the groomee. Further, social scratch is not elicited by the groomee: there is no specific presentation for social scratch of an itchy shoulder to the groomer, and no reciprocity. So, why scratch another?

If being socially scratched is pleasant (which seems likely), then scratching another may be empathetic. That is, the scratcher must recall from being scratched herself that it feels good, and so by second-order intentionality ('I know that you know') can put herself in the groomee's place. Thus, to engage in social scratch is to show intersubjectivity, a capacity hitherto restricted to human beings. Intersubjectivity occurs when two persons interact codependently in mutually satisfying sequences, e.g. a peek-a-boo game with a baby.

That such a simple behavioural pattern as social scratch exists in only one group suggests a behavioural 'mutation' that has caught on. It suggests that an innovative ape introduced the act and it spread through the group to become customary. We can never know this for sure, for by the time it was recognised in 1996, almost all of M-group already were doing it. Alternative explanations are unlikely: local people do not do the social scratch, so they could not have inadvertently modelled it for the apes. Ditto for researchers. There is no morphological evidence that the skin or hair of M-group chimpanzees differs from any other group, e.g. they have no exceptional dandruff, ectoparasites, skin conditions, etc.

The closest approximation to the social scratch is a behavioural variant done by the chimpanzees of the Ngogo community at Kibale (Nishida *et al.*, 2003). There, as part of social grooming, the groomer 'dabs' momentarily with the fingernails at the groomee's skin, using a single movement like a cat's scratch. Is this a form of social scratch, or another independently invented derivative of social grooming? Attention paid to the details of motor patterns in other populations might yield other variations on this theme. But social scratch is not mutual grooming, nor possibly even reciprocal. What is needed is a clear example of cultural grooming done mutually.

GROOMING HAND-CLASP

In January 1975, the first working visit occurred between the two oldest chimpanzee field sites – Gombe and Mahale – in western Tanzania. (Previous visits back and forth had been social or prospecting, but no data were collected.) Caroline Tutin and I from Gombe were hosted by Junichiro Itani at Mahale. On our first day out, near Myako camp, we saw K-group's chimpanzees do a striking behavioural pattern that was completely new to us. (At that point, we had studied Gombe's chimpanzees since October, 1972.) Yet, as recounted in Chapter 1, when we remarked on this the same evening to Professor Itani, he was unimpressed. Essentially, he said, don't all chimpanzees do this? At that point, I realised that all of us had probably been labouring under a simple illusion: that chimpanzee social life was the same everywhere. (Goodall (1973) had called attention to differences in material culture across populations, and to subcultural differences within Gombe, but at that point she had never seen wild chimpanzees elsewhere in Africa. On this basis, we described our preliminary findings on the grooming hand-clasp as a social custom (McGrew & Tutin, 1978).

What we saw were two chimpanzees sitting on the ground facing one another, engaged in mutual grooming (see Figures 1.1 (p. 2) and 8.1). At some point, at the start or during a bout, each fully

FIGURE 8.1 Mutual grooming by two adult males at Gombe. Each grooms the other's upper arm.

extended one arm (left or right) overhead and clasped the other's hand. This created a sort of A-frame postural configuration that revealed the armpit of the raised limb, which was then groomed by the other's opposite hand. This lasted usually for less than a minute, at which point they either switched hands or resumed ordinary mutual grooming. The two participants were in perfect symmetry, mirror images of one another.

In the preliminary study, grooming hand-clasp pairs were always of mixed sex, and only adults took part. In follow-up studies of greater depth, McGrew et al. (2001) and Nakamura (2002) showed that both sexes took part equally and that juveniles occasionally joined in. (The real constraint is the length of the arm: for full extension, both partners have to be about the same size.)

The grooming hand-clasp has a chequered distribution across Africa (Whiten et al., 2001). In Tanzania, it is customary at Mahale but absent at Gombe. In Uganda, it is customary at Kibale but absent at Budongo. In west Africa, it is habitual at Taï but absent at Bossou.

In seeking the origin of the grooming hand-clasp, one might start with a related behavioural pattern: branch-clasp grooming (McGrew et al., 2001). It resembles greatly the grooming hand-clasp in all ways bar one: instead of clasping one another's hands, the two mutual groomers each clasp an overhead branch (see Figure 8.2). Thus, unlike hand-clasp grooming, branch-clasp grooming requires overhead vegetation at the right height (1–2 m). This may be an ecological constraint, especially if chimpanzees prefer to group in open spaces lacking the right branches. This seems to be the case at both Mahale and Kibale (Kanyawara), where chimpanzees commonly groom while resting on trails, which lack low overhead vegetation. However, no one has yet studied branch-clasp grooming, so no data are available to test these ideas, although the pattern is likely to be a chimpanzee-universal (Whiten et al., 2001).

Usefully, as discussed earlier, the grooming hand-clasp also has now turned up in a captive group of chimpanzees (de Waal & Seres, 1997). Having first been seen in 1992, shown by Georgia, at the field

FIGURE 8.2 Branch-clasp grooming by two adult males at Gombe. Each clasps an overhead branch with the left hand and grooms the other's armpit with the right hand. Independently, an adolescent male on the right unilaterally grooms one of the adults.

station of the Yerkes Primate Center, it has spread gradually to more and more members of the group, starting with Georgia's offspring and close associates (Minter, unpublished data). (This spontaneous and continuing dissemination is in marked contrast to the failure to induce social learning of contrived gestures in the same group, Tomasello *et al.*, 1994.)

The second revelation about the grooming hand-clasp came only after we returned from a field season at Mahale in 1996. In viewing an extensive set of videotapes and slides, we realised that M-group's use of the hand-clasp differed from that of K-group (McGrew *et al.*, 2001). While K-group's apes usually clasped hands palm-to-palm (the 'classic' style), M-group's apes never did so. Instead, M-group's clasping was sloppy, sometimes even to the point of no clasping, just perfunctory touching of wrists (see Figure 8.3). Often, one chimpanzee would clasp the other's wrist, while that hand hung loose or rested against the clasper's arm. The M-style clasping was mostly nonsymmetrical, and often the arms were not extended fully. All in all, M-group's clasping seemed desultory and careless, by comparison to K-group's tendency to stick to form.

However, the asymmetry in performance by M-group yielded unsuspected additional information: in a subset of cases ($n = 8$) we could discern which participant supported the combined weight of the upright arms, because one participant rested his hand on the other's. In all cases, the supporter of the grooming hand-clasp was subordinate in social rank to the other, who did none of the work of maintaining the clasp.

Nakamura (2002) went on to test these ideas in greater detail, using behavioural data rather than archival images. The results are congruent. In another study, Nakamura and Uehara (2003) used a much larger photographic archive to seek to replicate the pilot study of McGrew *et al.* (2001), with regard to intergroup differences. Nakamura and Uehara found that the rate of palm-to-palm grooming in K-group was about 8 times more frequent than in M-group (42 versus 5 per cent). All four cases in M-group of palm-to-palm-style clasping

FIGURE 8.3 Grooming hand-clasp, non-palm-to-palm, at Mahale. Dominant individual in foreground (showing back) rests hand on wrist of subordinate (facing front) in background.

were done by one of the two members (Gwekulo, adult female) who had grown up in K-group before emigrating to M-group.

Variation in hand-clasp grooming is the best example yet of inter and intrapopulational variation in arbitrary social customs in apes. One awaits with keen anticipation similar analyses in populations other than Mahale.

CROSS-SPECIES SOCIAL TRADITIONS

At most field sites where chimpanzees are studied, there are sympatric baboons: olive baboons at Gombe, yellow baboons at Mahale, Guinea baboons at Assirik, mandrills at Lopé. Typically, these two taxa compete as the largest primates on the scene, both being omnivorous and opportunistic. Thus, it is not uncommon for baboons, with their more robust digestion, to take unripe fruits before they are ripe enough for chimpanzee tastes. Besides this usual scramble competition, there is even contest competition, e.g. when an oil palm is in fruit, its crown may be fought over fiercely. More spectacularly, they may pirate one another's most prized foods, e.g. bushbuck fawns (Morris & Goodall, 1977). Most of the time at most sites these two taxa live parallel lives without behavioural interaction, but there is one exception.

When Jane Goodall sought to accelerate the habituation of the Gombe chimpanzees, she began to offer bananas as bait in a clearing in the forest (Goodall, 1971). She was prompted to do this when wild chimpanzees burgled her tent at camp and stole bananas. Despite the drawbacks (Wrangham, 1974), the strategy worked and the apes became fully habituated; Gombe went on to become the most famous field site in Western primatology. However, there were two unexpected by-products to the process, both of which are informative.

The early protocols that took the bananas from the storage hut to the apes were nonselective. At first, the fruits were just strewn about on the ground, available to any creature bold enough to take them. Baboons were tempted, and succumbed, and so they were accidentally habituated, too, as the two species contested these treats. In the process, the chimpanzees and baboons were brought together for hours daily, in close proximity. (This was unnatural, but no more so than luring birds to a back-garden feeder, or inducing monkeys to enter hot springs by floating apples there.)

Two things changed in chimpanzee–baboon relations. First, the apes began to hunt, kill, and eat the monkeys, apparently being drawn irresistibly to baboon infants up close and vulnerable in the open feeding area. Before provisioning began, baboons had been a minor food-item in the predatory diet of the chimpanzees, far behind red colobus monkeys and even ungulate prey. During the years of heavy banana-provisioning, however, baboons became the predominate prey; after banana-feeding was stopped, the baboons returned to their minor position in the apes' dietary repertoire, which has held for the last 25 years (Goodall, 1986b). Nothing forced the chimpanzees to start eating baboons; the apes just made the most of a new development and exploited it as long as it was available. In other words, there was a fad for baboon-eating, unique amongst all wild chimpanzees studied anywhere in Africa.

The second change was even more unpredictable: the two species began to engage in affiliative social interaction (see Figure 8.4). First were locomotor play, rough-and-tumble play, and object play among youngsters. Pairs and groups would chase and flee, wrestle and mock-bite, tug-of-war with vegetation. Then came sociosexual exploration and interaction, with presenting and mounting (but apparently not intromission). The chimpanzees displayed at and teased the baboons, with much flailing of vegetation and throwing of objects. In quieter moments, there was even unilateral social grooming across the species. All of this was led by youngsters, especially juveniles and adolescents, as peers, with adults of both species playing almost no part (see Figures 8.5 and 8.6).

Why did this emerge? It is unlikely to be a direct result of eating lots of digestible, high-calorie foodstuffs. Chimpanzees and baboons raid crops, including banana plantations, in lots of places, but do not play with one another. The key would seem to be the context: the primates were drawn daily to the same hospitable location, where

FIGURE 8.4 Adolescent female olive baboon at Gombe presents to two adolescent male chimpanzees. Baboon has relaxed facial expression while two apes have play face. Photograph by Caroline Tutin.

having sated their energy requirements, they had lots of leisure time, which the adults used to sleep and groom. The youngsters had time on their hands, and sometimes for chimpanzees a shortage or even absence of same-species playmates. (One of the costs of a fission–fusion social system is that if your mother often ranges alone, you spend lots of time without peers (McGrew, unpublished data).) This may explain why the apes so often were the initiators of play bouts.

Decades later, the former banana-feeding area remains a well-drained breezy hillside open space. However, unlike the predation on baboons that died out, the play between monkey and ape continues, although neither species lounges around in the place these days. Now, youngsters of the two species play whenever they meet, most obviously to the observer on the trails used by both primates and humans. (This description must remain qualitative, because no one has studied

FIGURE 8.5 Adult female chimpanzee sits on stone and cradles her infant son in her arms. Juvenile female grooms the mother; shortly thereafter the mother released the infant to play with the juvenile. Photograph by Caroline Tutin.

systematically mixed-species socialising in the 40+ years of Gombe's primatological history.) So, now there is a multigenerational mutual tradition, likely to be in its third generation for chimpanzees and probably treble that for baboons.

FIGURE 8.6 Infant male peers at leaf being groomed by juvenile female chimpanzee at Gombe.

Like the predation, this chimpanzee–baboon overlapping social life is unique to Gombe and unnatural in origin. (Having said this, it is *not* unique, however: chimpanzees at Bossou capture and play with hyraxes (Hirata *et al.*, 2001b).) But that is no different in principle from unlikely cultural fusions across cultures such as Fijian rugby, Jamaican cricket, or American sushi bars. Vive la syncretism!

Lest the impression be given that chimpanzees are unique or even unusual in regard to social customs, recall the variety of 'games' played by white-faced capuchin monkeys (see Chapter 5). These seemingly arbitrary, even bizarre (sucking another's toes, sticking a finger up another's nose for minutes on end) interactions exceed in notability anything yet described for wild chimpanzees. Perhaps the closest social activity to this in apes is the rain dance (see above) with its slow-motion line dance, as seen in all but one of the habituated populations of wild chimpanzees (Whiten *et al.*, 2001).

A variation on the theme of falling water is another unique behavioural pattern of chimpanzees from Gombe: the waterfall display (Goodall, 1986b). In the main valley of the Kasakela community's range is a free-falling cataract at least 10 m high. It crashes down between rocky crags on either side to a plunge pool below. The constantly moist clifflike surfaces are verdant with vegetation, from ferns and mosses to woody lianas as thick as a garden hose. Through wet and dry seasons, the flow rises and falls, but the waterfall is always there, the most sensorially impressive feature of the landscape. Other chimpanzee populations (e.g. Mahale) have waterfalls, but they do not show the display (Nishida, 1980a). Instead, they toss boulders into pools to produce huge splashes.

Gombe's chimpanzees respond to the waterfall whether with others or when alone. The solitary response is notable particularly, for, unlike most displays, there is no one else present to display to! Adult males scale the lianas and whip them back and forth, swaying in mid-air and bouncing off the crags. Tarzan-like, they swing from vine to vine in a bout of gymnastics that is exhausting just to watch, as they ascend and descend. The concentration and persistence is total, for one mistake could mean injury or death on the rocks below. What is going on, especially when this is done alone, with no audience? Perhaps it is the ape's equivalent of a workout at the health club, or a practice session for a new display routine, or a chance to blow off steam. It is supremely situational: most wild chimpanzees, especially in rainforest, have never seen a waterfall, and so can never do it. But, like all phenomena, the waterfall display demands explanation.

CULTURAL LIFE

What do the following have in common: custom, institution, moré, rite/ritual, role/status, and taboo? All are societal norms or conventions, i.e. mental constructs that label acts endowed with social significance, or meaning. They are more than raw behaviour on an individual basis. They are acts that exist only in a group, implicitly by recognition if not explicitly by assertion. All of these ideas have in common their collectivity.

Sociocultural anthropologists once dealt with these conventions as their stock and trade. Their explanation was a challenge, and their diversity a testimony to apparently unlimited human imagination. When these conventions were found everywhere in all known cultures (e.g. marriage, incest avoidance, etc.) their universality was seen to be a clue to human nature. Human nature is what binds us together as a species, despite our infinite variety. These days, such thinking is thought to be obsolete, even embarrassing in its objectivity and specificity. Modern sociocultural anthropology with its discourses, voices, and texts seems to have no need of attending to what people actually do. It seems to be enough to pretend to read the mind of the other and to speculate upon the impressions gained. (For polemical treatment of these themes, see Kuznar, 1997; Sidky, 2003.)

Here I return to these traditional ideas for one simple reason: empirical utility. Winick (1960) defined these terms in ways that can be framed as testable hypotheses (see Table 8.1). If key elements of human culture thus can be operationalised, then one can ask if any of them occur in ape culture. The exercise may be old-fashioned, but if it is productive, who cares? (Goodall, 1982, 1986a attempted similar analysis, but less systematically and less specifically related to sociocultural anthropological ideas.)

(1) Custom

The grooming hand-clasp described above seems to qualify as a custom by Winick's (1960) definition (see Figure 8.1). It was first seen decades ago by observers at Mahale, and shows group-specific characteristics, e.g. differences between neighbouring K- and M-groups. However, we cannot say whether or not a religious precept underlies the grooming hand-clasp. If religion invokes the supernatural, and we lack an operational definition of the supernatural – which by definition is

Table 8.1 *Traditional, operationalised definitions of some terms signifying human culture, adapted from Winick (1960).*

Term	Definition
Custom	Group behavioural pattern, established by tradition, or contemporary social habits, or religious precepts. Enforced by social disapproval of their violation, but lack state coercion or sanctions of morés
Institution	Enduring cluster of social usages that are complex, integrated patterns of behaviour exerting social control
Moré	Accepted, traditional, and slowly changing behavioural patterns
Rite/ritual	Set of acts, involving supernatural, in traditionally established sequence. Often stem from daily life
Role/status	Dynamic aspect of status, which is comparative prestige rank in a community. Collection of associated rights and duties. Ascribed status is inherited, independent of individual capacities. Achieved status arises from special qualities, through individual effort
Taboo	Prohibition, which if violated, leads to automatic penalty inflicted by supernatural

beyond the bounds of science – we are stymied. This is similar to the ethnographer who records the funerary cremation of a dead person in a traditional human culture. The custom may be transcendentally religious – to send a person to her/his eternal rest – or may be community public health – to be rid of a potential source of illness. The former may be inferred on grounds of verbal report by the survivors of supernatural elements. Chimpanzees cannot tell us whether religion motivates their acts, but even if they could, it would not be verifiable.

Similarly, testing the prospect of enforcement by social disapproval of a violation of a custom is not easy. Violation of a (statistical) norm and enforcement by consequences are definable operationally (i.e. capable of being measured behaviourally) but what about disapproval (which implies a moral judgement)? As with our fellow humans, one can try only to draw sound inferences, based on their acts. De Waal (1991) gave examples from a captive colony of chimpanzees, but what do we know from nature? Infanticide presents a striking example.

Infant-killing is well known from chimpanzees of several populations (Bygott, 1972, Goodall, 1977; Nishida *et al.*, 1979, Arcadi & Wrangham, 1999; Watts & Mitani, 2002b). The general interpretation is that male killing of infants is more common as a straightforward reproductive strategy of killing the offspring of competitors and often done by immigrants (van Schaik, 2000). Female killing of offspring is less common as a strategy for eliminating competitors of one's offspring and is always done within the group (Goodall, 1977). But what about males who kill infants from their own group? Lacking certainty of paternity, especially in the absence of consortship, they could be killing their own offspring. This seems foolishly maladaptive.

Such intragroup killing occurs at Mahale (Nishida & Kawanaka, 1985) and may be explicable evolutionarily, as a custom. Normally, when a female immigrates into a community, she is welcomed by the males. She joins the group, travels in their core range, and bears young fathered by the group's males. These offspring grow-up unmolested. However, when a female fails to integrate fully into the group, and continues to range on its periphery – where she is accessible to neighbouring males – she risks losing her offspring to infanticide (Hamai *et al.*, 1992). It is the problem of the 'cuckoo in the nest'. Thus, the males are enforcing their disapproval of her violation of the norm of an immigrant showing sexual allegiance to her new community. Punishment leads to behaviour modification. Furthermore, when within-group infanticide occurs, it is male-biased, towards sons. Why?

Perhaps because in a species where males never leave their natal community and grow up to join the co-operating male club, it is important that these sons be genetic kin. Daughters will grow up to emigrate, to have their offspring elsewhere, and so are not a threat to the long-term integrity of the group, so they are spared.

Does this sound far-fetched? It is certainly controversial (Sommer, 2000). If this account seems anthropomorphic, try the following little test: substitute human for chimpanzee, leaving all else the same, then see if it becomes more credible. If so, then ask yourself if the evidence for infanticide in humans is more substantial than the evidence for apes. (Hint: with a few exceptions (cf. Schiefenhövel, 1989).) We are willing to accept exotic conventions as customs in human beings, almost without thinking, yet impose higher and narrower goalposts for other species. Such speciesism is understandable but regrettable.

(2) Institution

Whereas a moré is a behavioural pattern, an institution is a complex of patterns that are linked functionally. It is their integration that yields the social control that is crucial to Winick's (1960) definition (see Table 8.1). The most obvious and all-embracing institution is the social (or dominance) hierarchy.

Many species of animals have such hierarchies. The phenomenon was first described in domestic fowl, whereby each hen in a flock can be ranked from top to bottom based on aggression (or at least assertiveness) and submission over access to a prized incentive, usually food. Such a hierarchy is a predictable structure, e.g. if A dominates B, and B dominates C, then we can expect A to dominate C. But this social structure may be based entirely on individual, dyadic relationships, such that the inclusive hierarchy is only a by-product.

The chimpanzee hierarchy suggests a collective awareness that goes beyond individuality. Consider sex: in a one-to-one encounter every adult male is dominant to every adult female, with no exceptions. Even the biggest, healthiest prime female submits by pant-grunting to the smallest, weakest ageing male. Thus, it is not sexual dimorphism, but something that begins to look like gender. (Working together, several females can sometimes see off a single low-ranking male, but even this is notable, and he never pant-grunts to them.) This suggests that the female is reacting not to the individual male but to the sanctions that co-operating males can and do impose upon subordinates. No female can defeat a set of males who gang up on her.

Social rank prevails throughout daily life, whether over access to resources (e.g. food, water, nest site, even shade) or to social partners. Sometimes, threats to the social order are dealt with collectively: when a young adult male at Mahale began to challenge his superiors, he was attacked severely by the group, even the females (Nishida et al., 1995). A similar social punishment at Ngogo was fatal, which is the ultimate social control.

Another institution is the consortship, first described at Gombe (Tutin, 1979). A male and a female, both sexually mature, retire to seclusion from the rest of the group, for hours, days, or even weeks. They do so by sneaking away silently to the periphery of the group's territory, where they are less likely to be disturbed. This is risky, for it can put them in harm's way, vulnerable to attack by neighbouring patrols. Why take this time, effort, and risk?

The answer seems to lie in mate choice: by removing themselves from the domination of the alpha male (who can monopolise a female) and from the promiscuous mating of the males en masse, they are choosing to reproduce, or at least increase the chances of conception, with one another. (This need not be a conscious decision, any more than was the leap in births 9 months after the famous New York City power failure.) Intentional or not, conceptions on consortships occur much more often than from promiscuous mating, especially when compared as a rate: many fewer copulations are needed to produce a pregnancy during consortship (Tutin, 1979; Wallis, 1997).

This institution goes beyond a pair's co-operative mating strategy because of the way other group members deal with it. When a male seeks to sequester a sexually swollen female, he must do so despite the vigilance of the other males. If a female cries out in refusal to be purloined, this guarantees its failure, as the rival males come running, and the chance is lost. Furthermore, the return of the consorting couple to the group is fraught with risk, for they may be attacked upon arrival by the apparently infuriated and hoodwinked males. Also, consorting is more than a sexual liaison; sometimes the female is not even sexually receptive and no mating occurs over weeks. Sometimes the latest-born offspring of the female goes along, making it a domestic threesome.

Not all populations of wild chimpanzees show the institution of consortship, but it is hard to be sure, as it is so surreptitious. In some places, consortship may be precluded by a small home range: no place to go into seclusion, or a group size so large that it is practically impossible for a couple to slip away without being seen. In other populations (Taï) the institution seems not to have emerged.

(3) Moré

A norm is a statistical predominance. If the species hunted most commonly by chimpanzees is the red colobus monkey, then it is their normative prey. A moré goes further, as it must include collectivity rather than just the summed acts of individuals. Thus, co-operative hunting of red colobus is a moré in some chimpanzee groups, but is lacking in others. It does not matter if the presence or absence of hunting collaboration is influenced environmentally or even dictated, any more than it matters that the bison-based culture of the Comanche Indians varies when there are no buffalo. Social norms endure and evolve, just as they did when the coming of the horse via the Spanish transformed the bison-hunters.

If chimpanzee foraging is thought to be tied to too many ecological factors, consider male patrolling (Watts & Mitani, 2001). In this collective enterprise, male chimpanzees (mostly adults) travel to the edge of the group's territory, seeking neighbours or their signs at the boundary. They travel silently on the ground in bunched concentration, pausing often to listen intently or to sniff abandoned nests. Their nervousness is indicated by bristled hair and silent grimaces, and is well justified. If greatly outnumbered, an individual may be injured or even killed by the neighbours (Goodall *et al.*, 1979; Manson & Wrangham, 1991; Wrangham, 1999; Muller, 2002). The key is over-whelming numerical predominance, so that a neighbouring individual can be pinned to the ground by some members of the patrol while others pummel him with the hands and feet and slash him with their canine teeth. In such a fatal skirmish, the victim may be battered, eviscerated, and even castrated. The extent of such patrolling varies from habitual in some populations of chimpanzees (Gombe), to apparently absent in others that have no neighbours (Bossou).

Patrolling is a special case of a species-typical moré of chimpanzee social organisation: fission–fusion. No community of chimpanzees goes about its daily life as a troop (i.e. as a mixed-sex, -age, and -kin group of constant integrity) though this is usual for nonhuman primates, even closely related species, e.g. the gorilla.

Instead, chimpanzees range in parties, which are ever-changing subgroups. In a single day, an individual may be both solitary and social, and the latter may vary from a handful to most of the community. Moreover, the function of these parties may vary: mothers with youngsters may form a nursery party, just as males may go patrolling, or a couple may slip away on a sexual consortship. Sometimes, such an aggregation is driven environmentally: a bumper fruiting crop of *Garcinia* at Mahale or *Spondias* at Fongoli will bring apes from throughout the group's range to feast together, with much commotion. On the other hand, when a major fruiting species 'fails' in a season, the same community may be reduced to parties of two or three scrimping individuals, widely dispersed.

Party sizes may vary with social variables: a sexually swollen (and so receptive) female increases the party size. Ditto for party composition: a female may readily accede to a consortship with one male suitor, while vigorously rejecting the advances of another would-be consorter. Even the milieu of a party may vary, and so affect its duration: a party containing the alpha male and his main rival exudes tension and is likely to split sooner than one containing only old friends. Fission–fusion social structure is not unique to chimpanzees, but in no other species except human beings is it so varied and multifunctional.

(4) Rite/ritual

Rite or ritual entails a traditional sequence of acts, embued with the supernatural. The latter remains a problem heuristically here, as it did above. Candidates for rituals in chimpanzee ethnography include the waterfall display or the raindance at Gombe, as described above. The raindance is collective commotion, obviously stimulated by the impressive power of thunder, lightning, and deluge. If we humans can be in awe of the weather, even when we know what it is and how it works, imagine what a thunderclap can do to an ape. It is notable that instead of cowering from the elements, they display at them.

The waterfall display is more enigmatic. Why perform a display when there is no audience? It takes time and energy and entails risk. Moreover, the waterfall is constant; it never moves, always looks and sounds the same, so why not habituate to the stimulus? Perhaps this predictable power, confronted voluntarily, presents a chance for challenging exertion, or to practice display performance with adrenaline pumping. (Can anyone not stand at the base of a waterfall without their pulse racing?) Or perhaps it is a ritual of respect or abnegation to the god of water. This is sheer speculation, unlikely ever to be tested.

What about specific types of ritual, such as rites of passage? It seems that to become a fully accepted member of the adult male community, a young male chimpanzee must complete a task. He must dominate every adult female, one-to-one. This he does systematically: first he starts to tease females, by little threats and incipient displays, such as throwing objects at them. This causes them to react, but he stops and retreats before they can get at him. Eventually, this teasing turns to real challenges, starting with the lowest-ranking female, whom he threatens and pummels until she pant-grunts to him.

The vocalisation is an acknowledgement of subordinance, so then he moves on to the next female and repeats the sequence. This intimidation must include even his mother, whose relationship to her son is changed for ever. To complete the process of asserting dominance over all adult females in the community takes months but, once achieved, he has adult male status.

(5) Role/status

The topic of role was covered earlier for the splendid division of labour practised by group-hunting chimpanzees at Taï (Boesch, 2002). One can conceive of drivers, blockers, ambushers, and attackers almost as pieces on a chessboard. One might object that this is so far known only from the Taï population and wish for something found more generally across chimpanzee societies. Consider the following chimpanzee universal: at the top of every social hierarchy known in the species, in captivity or nature, is a single alpha male. Winick (1960) emphasizes the dynamic aspect of status, and the prestige that goes to the holder. It is a bit like the heavyweight boxing championship: a contender works his way up, achieves the crown, defends it against challengers, eventually is dethroned, and sinks into obscurity. There are a few ape exceptions, of ex-alphas staying high-ranking (Kalunde at Mahale) or even recapturing the crown (Goblin at Gombe), but these are rare.

More interesting than the trajectory is how alpha rank is attained and how it is managed, for here the potentials and limits of the role are played out (Kawanaka, 1990). First, being alpha is much more than getting first bite of the food or first choice of the females. An alpha male may break up a grooming bout between one of his allies and a rival, apparently in order to keep them from becoming too friendly (Nishida, 1983).

It is not necessarily the biggest or toughest male who becomes alpha. It may be the most inventive, e.g. Mike used the din from drumming on paraffin tins to cow and dumbfound his younger and stronger opponents (Goodall, 1971). It may be the best-connected, e.g. three (Figan, Freud, Frodo) of the last six (Goblin, Wilkie, Gimble) alpha males at Gombe came from one matriline, the biggest and most dominant one in the community. Some alphas achieve their top rank by acting alone, while others rely on a key ally, who may take the beta rank. In Mahale's M-group, Kalunde remained a kingmaker long after losing his alpha rank and into old age, by skilful (and sometimes fickle) alliance tactics.

In maintaining alpha status, style is obvious. These exemplify the right and duties of Winick's (1960) definition. Some alphas are nervous bullies, and others calm and tolerant. Some hold on by strength and vigour (easily assessed from their regular agonistic display), others by guile. An alpha male may use nonaggressive means to maintain his position, e.g. selective sharing of a prized resource: meat (Nishida et al., 1992). At Mahale, Ntologi was Machiavellian in rewarding allies, paying off potential trouble-makers and shutting out rivals. His successor, Nsaba, took possession of every monkey carcass and shared the meat widely with females, but with only one male, his close ally Kalunde. At Gombe, Freud was an assiduous groomer of his companions, and notably calm and nonaggressive. His successor, Frodo, never groomed anyone but instead intimidated them by brute strength. The atmosphere in the group seems to depend on the personality and performance of the alpha, but this has never been tested systematically. It is likely that more easygoing and secure alphas lead to larger, longer-lasting parties. Similarly, such an alpha is likely to be accompanied by more mothers and young infants, and by subordinates (e.g. orphans) than an aggressive alpha with a short fuse.

So far, almost all of the above relates to achieved status, in which the alpha position is gained by individual effort. The exception comes from matrilineal kinship: there seems to be a tendency, at least at Gombe, for alpha males to be individuals who have a maternal half-brother who is not much older. That elder brother backs up the challenging younger one in the struggle to the top, e.g. Hugh and Charlie, Faben and Figan (Goodall, 1986b). The exceptions who got to the top on their own (Goblin, Wilkie) had no brotherly pairs of similar age as competitors. Freud and Frodo reversed the order, with the elder brother becoming alpha first, followed by the younger. This fraternal alliance may be similar to ascribed status, though it is not based on direct inheritance.

(6) Taboo

When an act is proscribed, and its violation penalised by a supernatural agent, what is one to do to substantiate the existence of a taboo? Violation and punishment are at least observable, but prohibition and supernatural agency are more difficult. Prohibition in the form of physical restraint is easy enough to record, but what about implied threat? As for supernatural agency, divining it is the same problem as raised above, and just as intractable.

Or is it? Consider a superstition, which may take the form of a mild taboo: 'Don't walk on a crack, or you'll break your mother's back'. The violation (stepping on a crack in a pavement) and its consequence (spinal injury) are observable readily, but the prohibition is only verbal report and the supernatural agency is obscure. Clearly, it is not natural that treading on a substrate could cause injury to another, who need not even be present. Thus, supernatural agency is implied. As ethologists, we might suspect that a superstition exists by the unusual actions of the subject: stepping carefully and even awkwardly to make sure that the footfall is not on a crack. Such a deviation from normal behaviour would be obvious, as people do not typically stare at the substrate as they walk. So, in this case, anomaly in locomotion leads us to suspect something more, which might be congruent with verbal report (but only if the subject were willing to divulge it truthfully).

Consider another aspect of the chimpanzee's relation to water: at Gombe, chimpanzees have a phobia toward surface water (McGrew, 1992, p. 217). They avoid the edge of Lake Tanganyika for drinking, even if they walk along the beach. Normally, they leap or climb up and over streams, rather than wade across them, even when the water's depth is only a few centimetres (McGrew, 1977). Typically, an infant will climb onto and cling to its mother's back; then she will leap like a frog from one bank of the stream to the other. This behavioural duet is never seen otherwise. Even when walking along a path through the forest, a chimpanzee will detour around a shallow puddle rather than walk through it. Even when drinking, the normal pattern is to put only the pursed lips in contact with the water, and suck it in, with (for example) no use of the hands. All of this hydrophobic fastidiousness seems excessive, especially as it seems not to be a reluctance to get wet; chimpanzees make no attempt to take shelter when it rains.

There might be good reasons for avoiding surface water, as great apes cannot swim and so risk drowning, or water can be infective with the larval stages of intestinal parasites, but this does not stop other African apes. Both gorillas (Parnell & Buchanan-Smith, 2001) and bonobos (Thompson, 2002) spend hours daily foraging in swampy pools, sometimes up to their chests, with no ill-effects.

What about elsewhere? Chimpanzees at Mahale readily ford streams that look no different from those at Gombe (Nishida, 1980a). They hurl huge stones into pools as part of a soaking splash display (Nishida et al., 1999).

If we saw the compulsive reticence of Gombe's chimpanzees in a group of sympatric humans, we might be tempted to call it a taboo, based on the oddness of the actions. We would be likely to ask them about the meaning of surface water to them, and the penalty for failing to avoid the water. If our human informants produced a coherent verbal explanation, we would be likely to accord them a taboo. We cannot do that with apes, so we can only use the quack test. 'If it walks like a duck, looks like a duck, and quacks like a duck, then probably it is a duck!'

This chapter began with eavesdropping and ended with a water taboo. In between are description and discussion of the acts of chimpanzees that might be cultural. Methodologically, the constraints are great: sound inference is crucial. Sloppy speculation is tempting. All that we can do with chimpanzee behaviour is all that we can do with human behaviour: record it accurately and completely and persistently, then try to explain it logically.

The Azande of west Africa say that they seek the answer to an important question by asking the termites. They insert two sticks of different kinds of wood into a mound, and the one eaten by the insects provides the answer (Winick, 1960). Chimpanzees may do the same, so far as we know.

9 Lessons from cultural primatology

After 30 years of chasing wild chimpanzees and watching over the same period the emergence and growth of cultural primatology, I have a few opinions to offer. These views are seasoned with some humility; as mentioned earlier, our first paper on the subject (McGrew & Tutin, 1978) was squashed firmly by no less a luminary than Sherwood Washburn (Washburn & Benedict, 1979). Perhaps the best way to present conclusions is in the form of pithy epigrams, the basis for which has been given in the preceding eight chapters.

(1) Define culture as you wish, just make it operational
If there are as many definitions of culture as there are culturologists, then what are we to do? It seems pointless to pursue a consensus for the definition of nonhumans as studied by researchers from several disciplines, when students of human culture from only one discipline – anthropology – have never agreed on a definition (Kroeber & Kluckhohn, 1963). Furthermore, as there are two groups – the Humanists and the Universalists (see below) – at the opposing edges of the debate, there apparently is an unbridgeable gap: Humanists seem determined to define culture so as to exclude all species bar humans, and Universalists seem determined to define culture so as to include all species.

If there is no standard, all-purpose definition of culture that can be agreed on, then every practicioner can but try hard to be precise, explicit, comprehensive, and above all, operational in the definition chosen. Operational definitions are capable of empirical testing, i.e. their existence can be verified with validity and reliability. Thus, culture defined as the corpus of symbols making up the collective mind is not a useful definition, as both symbols and mind can only be conjured and never confirmed. Culture defined as a group's multiple traditions, socially learned and standardised, can be investigated in the usual reductionistic way of the scientific method, so long as each term (group, tradition, etc.) is delineated carefully. Thus, it is entirely reasonable to define 'culture' as tradition, but in doing so, some of the provocative issues raised earlier in this book are precluded, which seems a pity. The lesson here is to define the terms clearly and honestly.

(2) Labels are less important than content

Some people argue that the term *culture* is too loaded to be extended to other species than ourselves. They believe that the word is locked into referring to *human* culture, and so another label must be found for the nonhuman cases. So, *tradition* is redeployed to fill this gap for animals, leaving culture to us humans (Fragaszy, 2003).

Another way to fudge the issue is to use a 'not exactly' term. If what animals do is close-but-not-quite culture, then we can label it as 'culture.' Or we can use the wonderful array of prefixes available in English to qualify the phenomenon: 'proto' culture is incipient culture, 'pre' culture is a predecessor, 'sub' culture is a lesser version, 'quasi' culture has some similar features, and so on. The implication is that all lesser lights fall short of the real thing: Culture, as practised by *Homo sapiens*.

None of these weasel words is in itself objectionable. Some have been used, especially before about 20 years ago, with honest caution or doubt. Some have been used in a literal, evolutionary sense, as the ancestral form of culture, present in earlier hominids on their way to becoming fully human (Isaac, 1978). Much of the time, however, these qualified labels are the subject of intellectual laziness, because the key distinctions between culture and the near miss are *not* made clear. In science, we are obliged to use clear definitions, so as to be able to deal with candidate cases, and to classify them transparently. Nature may not be logical, but our definitions should be. The lesson here is that content is the key, and labelling is secondary.

(3) Culture as checklist – recipe for disappointment

As given in Chapter 4, one way to capture culture is to identify its essential features. Just as the inescapable ingredients of beer are barley, yeast, hops, and water, so in principle could a similar list of components be identified for culture. This reductionism yields a checklist, and if all the features are found to be present in a given case, then the phenomenon is verified. The most obvious such checklister was Hockett (1960) with his eighteen design figures for language and forty-five design features for culture (Hockett, 1973). Linguists still find this formulation helpful (Snowdon, 2001).

In cultural primatology, McGrew and Tutin's (1978) eight criteria for culture, or Whiten *et al.*'s (2003b) ten, contrasts serve the same function. So, what is the problem? First, unweighted or unranked checklists are simplistic, for some elements are more important than others. You can make beer without hops (just drink it soon) but not without yeast (it provides the ethanol). Second, elements rarely are as clear as they seem: you can also make beer with wheat, and from some other cereals, even if real beer is supposed to be from barley. Third, some items are neither clearly present nor absent: most brewers use a mix of different kinds of hop, so using just one kind of hop would be insufficient. Finally, the idea that a numerical score would be useful in some scalar way belies the heterogeneity of the list: if we had water and yeast, but not barley and hops, we would have two of four ingredients. Does that give us 50 per cent beer? If that sounds ridiculous, what if we had a creature that had social morés, teaching, and taboos, but lacked language, would it be 75 per cent cultural? The lesson here is that nothing as complex as culture (or language, intelligence, etc.) can ever be encapsulated in a list.

(4) Don't wait to know how before you ask what?, where?, when?, why?, etc.

If sociocultural anthropologists seeking to make sense of the rich tapestry of human diversity had waited until they knew the mechanisms underlying information transfer, we would still be waiting for the first monograph. We still have no idea what types of cultural transmission convey what proportion of knowledge from person to person, or generation to generation. Does teaching account for 1, 51, or 99 per cent of what we humans know? Professor Gerhard Wiens taught me German for a year in college, but did anyone teach me how to eat water-melon?

Worse, if we had waited until we knew the mechanisms, most of the traditional societies of the world would have disappeared before being studied, replaced by CD players and scented soap, leaving behind only fading oral history and enigmatic artefacts. Thank goodness no ethnographer ever delayed the quest for description and explanation of kula rings, potlatches, sky burials or menstrual huts, because it could not yet be said whether the knowledge of these conventions was based on imitation or emulation. No ethnographer balked at recording a custom because it could not be discerned under field conditions, whether it was taught actively or acquired by passive observational learning (Hewlett & Cavalli-Sforza, 1986).

Whiten and Ham (1992) outlined twelve channels by which information could be disseminated by the mimetic processes of cultural transmission. Byrne and Russon (1998) clarified and added more, and now Whiten *et al.* (2003a) have lengthened the list further in a critical review. In many cases, it may not matter which mechanism of information transfer is involved, and it is likely that more than one may be operating at once in real life, e.g. stimulus enhancement, response generalisation, and trial-and-error learning may all be involved in the acquisition of anvil use by wild chimpanzees (cf. Caldwell & Whiten, 2002). At any rate, we are unlikely to sort out the relative contribution of various mimetic processes to enculturation of the repertoire of patterns shown under uncontrolled field conditions by an individual ape. We can only hope that naturalistic experimentation can clarify the alternatives. Meanwhile, in nature, chimpanzee sons cannot learn to master the male charging display by watching their mothers, because their mothers do not do the display. As mothers spend most of their waking lives away from adult males, their sons must make the most of their chances when male models are available. Simple vertical cultural transmission will not do. This is only a problem for creatures with an extreme fission–fusion social system. Gorilla sons have no such problem as they accompany their fathers daily.

Finally, there are some cases where the *how* may be crucial: it seems unlikely that natural selection would reward individuals who altruistically expended time and effort teaching skills to the offspring of others. Teaching would seem to be evolutionarily sound only when it focuses on kin, where it might advantage them in development. Even more strange would be an individual teaching another who was likely to become a competitor. The sociobiology of teaching needs study. The lesson here is that culture functions according to who does what, when and where, whether or not we know how.

(5) Be sceptical of both Humanists and Universalists

In this scheme, Humanists are those scholars who seek to retain culture as uniquely human. They may do so by being prohuman or anti-nonhuman, or both. Social scientists, coming from disciplines fixated on human beings in historical time, seem to be the most likely to espouse Humanism when it comes to the possibility of chimpanzee culture. Natural scientists, with a worldview based on the cosmological continuity of evolutionary theory, seem to be the least likely to be Humanists. There are several predictable problems with a Humanist's approach to these issues.

One problem is the commitment to an arbitrary position of humanity over all. Implicitly, it smacks of arrogance, and explicitly, it has meant having to keep moving the goalposts. Goodall's (1964) reports of tool-making by chimpanzees demolished Man the Tool-maker. Similar erasures have followed new findings in primatology on tactical deception, logical reasoning, theory of mind, arithmetic, symbol use, etc. Anytime a Rubicon is declared, it is vulnerable to being crossed, and apes keep building bridges.

A second problem is history, or rather prehistory. Living humans and chimpanzees had a common ancestor about 6–7 million years ago, which is a long time for the parallel development of the two

evolutionary lines. How far back are we willing to grant cultural status in the human line? No one would object to the last 35 000 years,
given the wonderful images in the caves of southwestern Europe, e.g.
Lascaux, Altamira, Cosquer, and Chauvet. But what about before
that? As I write (12 June 2003), *Nature* has just published accounts
of 160 000-year-old modern humans that were scalped, apparently in
mortuary rites (Clark *et al.*, 2003). Does this count as culture? Presumably, yes. But what about cut marks 2.5 million years ago on ungulate long bones, as evidence of butchery (de Heinzelin *et al.*, 1999)?
These seem to qualify, too. If chimpanzees show percussive lithic
technology in cracking open nuts that is indistinguishable from early
hominid lithic percussive technology in cracking open long bones,
who has culture? Presumably, we must grant cultural status to both
extinct *Homo* and extant *Pan*. It is hard to be a Humanist when the
borderline of humanity is unclear.

Conversely, Universalists are those who seek to extend cultural
status as widely as possible. Sometimes the extension is logical and
hypothetical: if culture requires intelligence, and big brains yield intelligence, then one should look for culture in big-brained creatures.
Whether absolute (cetaceans) or relative (Corvidae (i.e. crows, jays,
ravens, etc.)), braininess is a good rationale. Sometimes the sought-
after extension is one of principle and open-mindedness: why should
not fish be cultural? Or, if one-celled animals can be facultatively
altruistic, then why not cultural as well? Thus, the contention that
apes are cultural can be seen as élitist. Universalists seem to believe
that for every question of comparative analysis, evolutionary continuity is the null hypothesis. Thus, until proven otherwise, gophers are
just as likely as gorillas to be cultural. Once social learning is shown,
then membership to the Culture Club automatically follows, as if
there were no difference between an octopus learning to avoid a threat
by watching another octopus do so, and an immigrant chimpanzee sit-
uationally altering her social gestures, depending on with whom she
grooms. Humanists fail to acknowledge that apes are now known to do
many things in nature that until recently were thought to be uniquely

human. Also, as shown in preceding chapters, apes do some complex things that are known in only one other species: *H. sapiens*. Universalists trivialise culture by dumbing down the idea (social learning equals culture) as if all creatures in principle were equal. This may be ethically so, but science seeks distinctions, and once found, these cannot be denied. The lesson here is to avoid extremists at the tail of any distribution.

(6) Don't buy the space shuttle argument

Culture in chimpanzees? Absurd! How can you possibly compare another species with us humans? After all, what animal ever constructed a space shuttle, invented algebra, composed a symphony, etc? Where is the animal equivalent to Shakespeare, Milton, Darwin, etc? (And that's just a list of dead Englishmen!) Put this way, the argument may make the cultural primatologist slink away back to the rainforest, to stick to sieving faecal samples.

There are at least two major problems with the 'space shuttle' argument (besides the obvious one that space shuttles are hazardous vehicles, not suited for most of us). One problem with it is that by these criteria, most human societies lack culture. Most human societies are nonliterate, minimally numerate, lack the wheel, rely on oral history, etc. Most traditional cultures before historical contact with outsiders could no more imagine an oxcart than a spaceship. They would have no need for algebra, even if they had a chance to meet and master it. Rather than deny culture to these fellow humans, perhaps we should be less ethnocentric in how we think of the phenomenon.

The other problem is that even in modern industrialised cultures with near-universal literacy and technology and higher education, how many of us ever invented or composed or devised *anything*, much less neurosurgery or cell phones with screens that change colours according to the user's mood? Most of us would fail on the 'space shuttle' criterion, so we can hardly look down our noses at the apes. Before we are tempted to class the chimpanzee sleeping platform as a primitive construction, how many of us have ever made any kind of shelter in our lives? The lesson here is to combine realistic self-knowledge and humility with a more inclusive appreciation of multicultural diversity.

(7) *Avoid anthropomorphism* and *anthropocentrism*

Anthropomorphism is the great sin of cross-species comparison, when one of the species being compared is *H. sapiens*. Of course, students of chimpanzees can become chimpomorphic if they seek to attribute chimpish abilities to other species, or to interpret the acts of other species in chimpish terms. This seems to be an occupational hazard for those who work with capuchin monkeys (see Visalberghi & McGrew (1997), or almost any article by Westergaard, e.g. that written in 1994).) Anyone doing such analyses risks being accused of overstepping bounds by exaggerating the abilities and performances of non-human species. In scientific terms, this is nonsense, as anthropomorphism is about explaining results. Explanations are a dime a dozen, and range from the wildly improbable to the highly likely. What matters are sound data that address falsifiable hypotheses. If we see a chimpanzee sitting up in his nest and staring out across Lake Tanganyika at a glorious sunset over the mountains of the Congo, we may interpret the act as contemplation of the meaning of life, perhaps brought on by the sight of the dying sun. Or, perhaps it is a compelling fixation on the sheer overwhelming beauty of the colours in the sky, prompted by some protoaesthetic sense. Such interpretations are untestable, but we can pay attention to when in the circadian cycle chimpanzees sit and stare at anything. If they restrict their concentration to displays of natural beauty (e.g. such as sunsets, sunrises, rainbows) then we have some data to work with, just as we do if we find that they just as often sit and stare at tree trunks.

How can we humans be anything other than anthropocentric? It seems natural that guppies should be guppicentric and echnidnas should be echidnacentric, and so on. Of course, the better we know another species, the more we can try to look at the world from its point of view. This is what makes long-term commitments to a lifetime of primatology focused on one group so valuable. With chimpanzees, it really is long-term: when Jane Goodall began research at Gombe in 1960, the female chimpanzee Fifi was about 4 years old. As I write this 43 years later, Fifi recently has produced her eighth offspring and is still going strong! When we try to make sense of chimpanzee life, we use ourselves as a starting-point of comparison. It may turn out that a better baseline would be the bottle-nosed dolphin, but we are who we are.

Of course, human culture is unique. So is human digestion. Equally, so are chimpanzee culture and digestion. The lesson is to draw telling and specific comparisons: it may be that baboons will tell us more about chimpanzee digestion, but for understanding chimpanzee culture, it is hard to think of a better point of comparative departure than ourselves. The lesson here is to be realistic as well as vigilant about points of comparison.

(8) Nature and nurture provide the context, but not the shackles

In seeking to advance the case for chimpanzee culture, it is tempting to try to eliminate environmental factors. That is, if one could show that standardised, socially learned variation in behaviour were independent of environmental variables, it would be more compelling evidence. Why? Because otherwise the ecological determinists always will be able to say that cultural variation is dictated by the environment. In some simple cases, of course, this is true: we cannot compare nut-cracking techniques across all populations of wild chimpanzees because some populations such as Assirik have no nuts to crack. Further, it is hard to compare the nut-cracking of two populations with access to nuts if one of them raids crops and the other hunts often. Other high-quality items in the diet – which are therefore part of the environment – can influence indirectly the cultural candidate: nut-cracking. Finally, no living creature ever is free of environmental influences: even with vocalisation – which anyone can utter anytime and anywhere – transmission is constrained by the physical environment. Calls carry further and more clearly in open woodland than in evergreen forest, given the physics of sound attenuation.

By the same token, cultural primatologists would like to eliminate the effects of genes. We would not like to learn that a population of apes that does not engage in nut-cracking lacks an enzyme for breaking down nut meat, and that enzyme is present in those that do practise nut-cracking. We would not like to learn that the chimpanzees of Uganda that lack skilled subsistence technology have innate sensory-motor deficits. We assume that this does not apply in both cases, because groups who are neighbours, and exchange genes, still maintain different customs (Boesch, 2003). But we do not know this for sure, and probably never will. The standard protocols of behavioural genetics (e.g. cross-fostering) are not likely to be done in nature for apes. Even in captivity, the longevity and slow reproductive rate of apes make such projects impractical, even if they were acceptable ethically.

Even if we could rule out direct environmental and genetic effects on cultural variation, we cannot rule out the feedback from cultural effects on the environment and gene pool. It is two-way traffic (Laland *et al.*, 2000). Hunting by chimpanzees affects the population structure of red colobus monkeys (Stanford, 1998a). Infanticide by chimpanzee males may well have selected for compliance in females, if females who failed to move to the core of the group's range suffered more infanticide and so had lower reproductive success.

But such constraints are not shackles. The glory of culture is its inventiveness. Culture means new solutions to old problems, and vice versa. No one has replicated Pryor *et al.*'s (1969) classic study of training for novelty in dolphins. Instead of rewarding their subjects for performance of particular learned responses, the trainers rewarded them only for doing something that they had never done before. Once the dolphins caught on, they added increasingly more intricate elements to behavioural routines, the complexity of which eventually exceeded the experimenters' capacity to keep up in recording them. I suspect that apes would show us the same creativity. When captive chimpanzees at a research centre began to escape by using poles as ladders, it was impressive but not totally surprising, as they had seen maintenance personnel use ladders (Menzel, 1972, 1973a). When they later began to jam sticks in the wall and to use them as pitons to scale the walls, this *was* surprising, as no humans had modelled that activity for the apes (McGrew *et al.*, 1975). This, by the way, is yet another form of ratcheting, the characteristic of culture still denied to nonhuman species by some commentators, e.g. Tomasello *et al.* (1993); Boyd and Richerson (1996); Alvard (2003).

The lesson here is that, just as with all human cultural activities, nature and nurture provide the inevitable backdrop. Culture does not transcend its players or their setting. Nature and nurture provide a sort of leash (Lumsden & Wilson, 1981), but culture cannot be shackled.

(9) Between-group variation and within-group similarity prove nothing

In considering the contributions of nature and nurture to culture, there is a tendency to equate nature with genetic–environmental determinism and nurture with cultural relativism. In this simplistic dichotomy, genetic–environmental means fixed and invariant, while cultural means plastic and variable. This is nonsense, for several reasons.

First, all sensible scientists are interactionists, and genes, environment and culture co-evolve (Lumsden & Wilson, 1981; Boyd & Richerson, 1985). For example, relaxation of lactose intolerance after weaning in human pastoralist groups is a biosocial problem (Durham, 1991). Natural selection acted on the genes for lactose production only in those groups that domesticated ungulates for dairy production. Similarly, diurnally active primates, including humans, normally give birth at night because such births were safer in the evolutionary past, but induced births (not surprisingly) are concentrated in the daytime for cultural convenience (Jolly, 1972).

In seeking to explain human behaviour, it is easy to focus on nurture and ignore nature, given rich diversity. But variation primarily can be determined genetically, and lack of variation can be determined culturally. Thus, we must also look at similarity across groups and variation within groups. Perfect pitch, as most obviously manifest in singing, is determined mostly genetically (Tramo, 2001). The rest of us can take singing lessons all our lives and never hope to achieve what someone else was born with.

To repeat a more flippant example, global consumption of sweetened carbonated, Cola-flavoured beverages mostly is determined culturally. Its invention and rapid spread are well known, but the prototypical Martian ethnographer arriving now might well add it to the list of human universals of greater antiquity (Brown, 1991). The same may be true of chimpanzee shelter-making; although a universal, it is cultural.

The lesson here is to presume nothing about the sources of behavioural variation, in our species or in others.

(10) Start with material culture, but don't stop there

Science is done more easily with concrete and particulate subject matter. Such objects can be measured, passed round, and curated, or if necessary deconstructed, both literally and figuratively. It is no surprise that the early ethnographers of wild chimpanzees published first their observations on material culture (see Goodall (1963, 1964); Boesch (1978); Sugiyama & Koman (1979b), etc.). Not only were these exciting contributions to the debate on Man the Toolmaker, but also there were other down-to-earth and good reasons: in the early days of any field study, when behavioural data are few and fleeting because the subjects are wary, at least they leave their tools behind. Thus, the form and function of elementary technology are available before the performance can be seen. Once recognised, material culture can be studied even on days when no subjects can be found. Many a field worker has been sustained by artefacts, when behavioural data were sparse.

Similarly, the material culture of individual subsistence and maintenance are the most accessible to researchers at the outset.

Tools used socially (e.g. weapons) require more information on who does what to whom. Repeated use of tools, including their transport, that were used by others before, may be common. In the Taï forest, suitable stone hammers for *Panda* nut-cracking are so rare that they 'circulate' from worksite to worksite (Boesch & Boesch, 1984a). On the other hand, collective use of tools is unknown in wild apes; they use no battering ram, or tug-of-war. The closest to the social use of an object by wild apes is something like the children's game of 'keep-away', where the possessor of an object is pursued until the pursuer gains possession, and then the roles are reversed. Wild chimpanzee juveniles play this game, but it has never been studied systematically. In captivity, chimpanzees using ladders may co-operate with one holding while another climbs, but for this, too, we lack good data (McGrew *et al.*, 1975; de Waal, 1982).

Typically, as populations become habituated to human observation, data on nonmaterial culture, usually as expressed in social interactions and relationships, come to the fore. Sometimes these are idiosyncrasies (e.g. pressing one's own nipples in nervousness (Nishida, 1994)) and at other times they are characteristic of the group, e.g. social scratching. These acts leave no artefacts and may last only seconds. Sometimes, behavioural patterns are taken for granted, even for years, because observers did not realise that these vary across groups or populations (e.g. grooming hand-clasp), or did not notice the finer points (e.g. use of leaf as platform to kill tiny ectoparasites), or assumed similarity of performance based on published descriptions, e.g. ant-dipping. Until ethnographers of chimpanzees take the time to exchange visits across study sites, more cultural nuances, both material and otherwise, will be missed. Finally, field researchers of chimpanzees should continue to seek connections between material and nonmaterial culture. Juvenile female chimpanzees sometimes seem to handle a piece of broken branch as if it were a baby. We can only appreciate this if we have knowledge of how babies are handled in that community, to see if the same behavioural patterns are employed. The piece of wood on its own may look like nothing special, but it is embued with meaning if it becomes a doll. It is likely that other chimpanzee toys have similar significance, but this needs further study (Ramsey & McGrew, 2004). The lesson here is to take advantage of materiality in chimpanzee culture, but not to be bound by it.

(11) Beware of anecdotes, but don't ignore them
Anecdotes are problematic, as discussed in Chapter 5, as five of the six possible reasons advanced for a single event will be wrong. A case of $n = 1$ could be the first instance of more to come (which is good) or it could be an accident, mistake, observer error, misattribution or hoax (which is bad). Everyone loves a story (another meaning for the word *anecdote*), but unique or rare events in science can be misleading or Earth-shattering. For chimpanzees, Beatty's (1951) two-paragraph account of nutcracking turned out to be prescient, to be followed by a host of reports decades later in Guinea and Ivory Coast. Bygott's (1972) account of a single event at Gombe of chimpanzees killing and eating a chimpanzee infant has since been seen in several other populations, as discussed above. Brewer and McGrew's (1990) description of a toolset to get honey has had both theoretical (Sugiyama, 1997) and empirical (Westergaard & Suomi, 1993) consequences.

On the other hand, some anecdotes have been provocative, but not yet been confirmed as customary, or even habitual. Boesch's (1991c) accounts of active tuition at Taï by adult females to their young have become legendary, in showing how teaching in nature might be done. Others are now looking hard for teaching in nature that has been overlooked before (Nakamura & Nishida, 2002). Matsuzawa's (1991) repositioned nut-cracking anvils at Bossou have been seen a handful of times, with no consistency. Finally, as seen in Chapter 5, Akko's knotted colobus skin necklace was seen only once, years ago, without repetition (McGrew & Marchant, 1998). Only time will tell.

Sarringhaus (unpublished data) analysed the citations of the ten most often-cited anecdotes in primatology for tool use and hunting. In both sets, anecdotes were often misused in their citation by others. Common problems were failing to mention that the event was an anecdote – leading to overestimation – or citing specific events as evidence for something broader – leading to overgeneralisation. It seems that scientists often let a good story get in the way of the truth.

The lesson here is: report everything of interest clearly and precisely in print, to spur others. Then, replicate, replicate, replicate. The plural of anecdote *is* data.

(12) *Raid sociocultural anthropology, selectively*

It might seem obvious to say that cultural primatologists have more to learn from cultural anthropologists than from anyone else. However, one suspects that most sociocultural anthropologists are not only uninterested in primatology, but also are resistant actively to such engagement. As cultural primatology grows in prominence, it raises problems for current, mainstream, anthropology.

First, cultural primatology is quantitative, positivist, reductionist, while presently fashionable anthropology is qualitative, relativistic, holistic. It is hard to exchange views when the epistemology is so disparate. Second, many of the ideas and methods that cultural primatology is discovering (and in some cases, re-inventing) and developing, are considered passé or even anathema in postmodern anthropology. Difussion theory apparently will explain the distribution of several chimpanzee subsistence activities (e.g. nut-cracking) but it is no longer considered useful in anthropology. Methodologically, cultural primatologists still collect data, which they analyse statistically, whenever possible. Cultural anthropologists engage in discourses with The Other or draw their knowledge from texts, which may be anything from verbatim transcripts of interviews to household utensils. There seems to be little overlap with cultural primatology, except with human behavioural ecology (Smith *et al.*, 2001; Alvard, 2003), which as a subfield is just about as marginalised from mainstream anthropology as is primatology. Finally, cultural primatology is a threat of sorts, for truly it pushes the boundaries of the discipline. If apes are admitted to the Culture Club, who then knows what other taxa will slip in through the door left ajar?

So, what can anthropology offer to cultural primatology, how-ever reluctantly? First, ethnographic methods can be adapted for pri-matology, by consulting any well-done monograph on a traditional society published before 1970. For example, the role of kinship is cru-cial to both fields (whether fictive as well as genetic kinship exists in chimpanzees remains to be seen; evidence may well be found in the soft grunts used in intimate communication, rather than from loud long calls). Second, data-handling techniques can be borrowed from ethnology. The high number of field sites now yielding data on wild chimpanzees is perilously close to information overload. We need both an archive for field data and reports, available online, as well as an atlas coded from that archive. The Chimpanzee Cultures Project is making a start (Whiten *et al.*, 2001). Many of the coding categories from the Human Relations Area File (HRAF) apply also to apes. The same anal-yses undertaken on human societies using a cross-cultural atlas can be done on chimpanzees, e.g. if group size and resource base are held constant, is the extent of boundary defence negatively correlated with territory size? Or is the key to confronting the neighbours the adult sex ratio, so that low male-to-female ratios lead to more boundary patrols? Why are nut-cracking and termite-fishing apparently mutu-ally exclusive? Is it merely environmental (e.g. based on annual rain-fall) or are the two foodstuffs nutritionally interchangeable, and so redundant? Many such comparative questions can be asked of large databases, and sociocultural anthropology has led the way in this area.

On the other hand, reliance on verbal texts elicited infor-mally from informants is not likely to help in cultural primatology. Participant observation is out of fashion in sociocultural anthropol-ogy, and the ambitious programme of cultural ethology called for by Lorenz (1977) and pioneered by Eibl-Eibesfeldt (1989) has never been implemented systematically. Sociocultural anthropologists who have kept track of the real-life acts and consequences of their hosts, then frankly reported the results (e.g. Chagnon, 1992) have fallen victim to political correctness. Here, cultural primatologists have much more in common with the behavioural ecologists, whether it is the link between hunting prowess and reproductive success in males (Kaplan & Hill, 1985), the trade-off choices between major subsistence strategies (Alvard, 2003), or the implications of the 'grandmother hypothesis', i.e. that older women may shift investment of time and effort from children to grandchildren, explaining the evolution of menopause (Hawkes *et al.*, 1998). All of these issues apply to chimpanzees, and the apes sometimes produce different solutions to similar problems. The lesson here is that cultural primatologists should persist, despite resistance, in scavenging the remains of scientific cultural anthropology for useful ideas and techniques.

(13) As with humans, there is diversity in chimpanzee culture
In the social sciences, there is a tendency for well-meaning scholars to assert that all humans are the same, or equal. Thus, it is politically incorrect to talk about differences based on sex (which must be sexism), age (ageism), race (racism), or even body composition (fatism). In sociocultural anthropology, all cultures are deemed of equal worth, and value judgements are discouraged. Words such as *progress* or *primitive* are tabooed, even if people use them (or their equivalents) to describe themselves or others. Thus, if one human culture has a vocabulary size that is ten times that of another, this is not thought to be important. Similarly, if one population lacks major technological features (e.g. fire-making, hafted tools), as with the Tasmanians, versus their counterparts on the Australian mainland, such discrepancies rarely make it into the anthropology textbooks.

In the natural sciences – apart from cross-populational differences in vocalisation in birds or cetaceans – little emphasis is given to what might be cultural differences across groups. When California sea lions use stone anvils to crack open molluscs, but Alaskan sea lions do not, it is assumed that some environmental difference accounts for the contrast, e.g. differences in potential food resources. The same argument applies to scrub jays, which in Florida live in extended families with adult helpers but in California live in the usual songbird

Table 9.1 *Cultural repertoires of eight populations of wild chimpanzees, from Whiten et al. (2001), Table 3.*

Measure	Field site							
	Gombe	*Tai	**Mahale	Bossou	Kibale	Budongo	Assirik	Lopé
Custom & habit	16	19	13	9	10	9	7	5
Presence	7	2	6	8	2	2	2	0
Total	23	21	19	17	12	11	9	5

*Redundant patterns of nut-hammer and food-pound combined
**Results for Mahale's K- and M-groups combined

pairs. We would never call the Alaskan sea lions primitive, or the Florida scrub jays advanced.

What about wild chimpanzees? Does their diversity resemble humans in this way? Table 9.1 is based on Whiten *et al.*'s (2001) data on the presence or absence of thirty-nine patterns deemed to show rudimentary cultural variation. In the table, these are presented from left (Gombe) to right (Lopé) in order of total variants recorded. Not surprisingly, the two study sites lacking habituated subjects (Assirik and Lopé) show the lowest totals. All behavioural patterns that do not leave behind material evidence would be hard to see there. Of the three sites to the left, there is no significant difference between Gombe, Taï, and Mahale. Bossou, with its single small community, tiny home range, disturbed habitat and crop-raiding, has fewer (nine) established variants. The real outliers, however, are the two Ugandan sites, Kibale and Budongo. They have low totals and show only *one* subsistence behavioural pattern – anvil use at Budongo – between them. So there seem to be real cultural differences across chimpanzee populations, even when degree of habituation is held constant. Ugandan chim-panzees seem to be primitive technologically, by comparison with their counterparts in east Africa and those further afield. The lesson here is that the pattern of chimpanzee behavioural diversity seems to be more like that of humans than of other animals.

(14) Engage with archaeologists – they have similar problems

As discussed above, researchers on unhabituated chimpanzees in nature often face the same set of problems as do archaeologists: absence of their subjects, and so of their behaviour. For archaeologists, these data are irretrievable, as the subjects are dead, which has led to enterprising and imaginative attempts to 'reconstruct' what went on in the past. (For a summary of such processual archaeology applied to human origins, see Schick & Toth, 1993.) For a specific example, consider cannibalism. We will never see extinct individuals ingest the flesh of other humans, so we must be most careful to exclude other alternatives: defleshing as mortuary rites, dismemberment as punishment for witchcraft, consumption by other species, etc. Ultimately, inference piled upon inference strengthens the case (White, 2001), although the one conclusive element (coprolite containing myoglobin) remains an anecdote (Marlar *et al.*, 2000).

In cultural primatology, we can have both the indirect (remnants of the victim in faeces as hair, bone, etc.) and the direct evidence through observation. With habituated chimpanzees, we can cross-validate one sort of evidence against the other: when an infant goes missing, we can check for remnants in others' faeces. We can sharpen up the indirect evidence (e.g. by attending to the teeth marks left on bones) and can try to distinguish ape tooth marks from those of carnivores (Tappen & Wrangham, 2000).

The lithic technology of living apes is minimal by comparison with even the earliest flaked technology assigned to early hominids. However, recent findings from both laboratory (Schick *et al.*, 1999) and field (Mercader *et al.*, 2002) make it less conclusive that all flaked stone in the archaeological record was produced by humans. It has been argued that apes with brains the size of australopithecines have all the cognitive abilities needed to produce the simplest flake stone tools (Wynn & McGrew, 1989) and there is a plausible evolutionary scenario to get from percussive (anvil) technology to the most basic knapping (Marchant & McGrew, 2004). There are many captive chimpanzees sitting bored and idle in captive facilities who could be tested on these problems, especially if they needed such tools to get their daily bread.

The sensible thing to do is to get primatologists and archaeologists beyond talking and into collaborative working, following the leads of Sept (1992, 1998), Joulian (1994, 1996), and Mercader *et al.* (2002). Primatologists should join in digs and in actualistic archaeological research; archaeologists should join primatologists at field sites and watch chimpanzee elementary technology in action. Most primatologists do not have a clue about how to map a work site. Most archaeologists can only speculate about how tools are acquired and transported, yet wild chimpanzees do this every day. Most primatologists underestimate what information can be derived from artefacts; most archaeologists underestimate the capacities of apes. The lesson is that these two disciplines are beautifully complementary, but because of their different literatures and training they rarely come together, yet this could so easily be remedied.

(15) Experiments are wonderful, but hard to do
There is a curious, revisionist idea floating about: that the only sure knowledge of culture is experimental, so that field studies are only useful as hypothesis-generators, to be tested in the laboratory (Galef, 2001; Laland & Hoppitt, 2003). In principle, it is true that causal and predictive analyses are better than correlational and postdictive ones, and the former usually are experimental and the latter observational. Equally true, however, is that some disciplines can never do crucial experiments, because of constraints in time (e.g. history) or space (e.g. astronomy), but this has not impeded the advance of knowledge.

These constraints do not apply to living humans, so why is sociocultural anthropology not experimental? The discipline started as natural history (opportunistic observation), moved on to ethnography (systematic description and classification), then ethnology (theory-driven, hypothesis-testing analysis), before drifting into intuitive speculation (postmodern discourse). Cultural anthropology did not move on to experimentation, because of logistical and ethical reservations. Cultural primatology has followed the same path of development, and now reaches the decision-point about experimentation.

In captivity, the number of experiments on cultural learning processes in chimpanzees has burgeoned (Whiten *et al.*, 2003a), and some of these are naturally inspired. In most cases, however, the models are human, the tasks artificial, and the settings contrived. Marshall-Pescini (unpublished data) has sought to model nut-cracking, ant-dipping, and termite-fishing, using (for example) an artificial fruit device tested on young wild-born apes. By going to Africa, she was able to use valid incentives (oil palm nuts) and sane subjects (wild-born, mother-reared orphans). This line of research shows promise but takes ingenuity and access to the right subjects.

In nature, apart from the early efforts of Kortlandt (1967), the only prolonged attempt at experimentation with wild subjects has been that of Matsuzawa and colleagues at Bossou (Matsuzawa, 1994). Their 'outdoor laboratory' is limited by natural constraints: trials cannot be scheduled nor subjects assigned randomly to treatment. Control conditions often cannot be managed, but baseline-treatment (before–after) comparisons are feasible, often with repeated trials. Perhaps more importantly, observation conditions are better and the frequency of behavioural performance is higher than in the forest.

The real problems with experimentation in field studies of cultural primatology are practical and ethical. In principle, one could follow Laland and Hoppitt's (2003) advice and swap ant-dipping chimpanzees between Gombe and Taï. If the two-handed Gombe pattern introduced by the immigrant caught on at Taï, or vice versa, then diffusion would be implicated. In practice, the experimental procedures of capture and release probably would damage the habituated status of both populations of apes, who would rightly be resistant to newcomers. The financial cost would be prohibitive, and after all that, there is no guarantee that the transplant of the community would survive, especially if the immigrant brought new pathogens. Imagine a Westerner magically transported to the remotest New Guinea Valley, with nothing but the clothes on his back. Would he be taken in and attended to?

The ethical debate is complex and sometimes obscure. In principle, some researchers would like each chimpanzee culture to remain pristine, or at least unaltered by immigrants. This sounds good, and one thinks of remote Amazonian peoples, resisting the globalising impact of outsiders, at least until the latter's seductive material goods become familiar. The analogy for chimpanzees is access to humankind's domesticated animals and plants, with the result usually of crop-raiding. One could imagine an ape imigré with a taste for sugar-cane who corrupted a host community, with fatal results. On the other hand, if human development threatens a chimpanzee population, are we to leave it to perish, rather than move the apes to safety? Ideally, transported ape refugees would occupy vacant areas, as with Rubondo Island, where chimpanzees released in the 1960s have now gone feral (Borner, 1985). To see if such introduced populations re-invent chimpanzee technology in a strange land is perhaps the biggest experiment of all.

One cannot do a real experiment in nature, by definition. Nature controlled is no longer nature, but there is a continuum. Naturalistic studies in captivity and experimentalistic studies in nature can narrow the gap. The lesson is that we can never hope to study culture by crucial experiment, but we can strengthen inferences by doing the best we can in the constrained circumstances described above.

(16) Social learning is the starting-point, not the end

Social learning does not equal culture; instead, social learning is a necessary but not sufficient condition for culture. You cannot have culture without it, but lots of social learning need not be culture, e.g. peeling the bark from your fishing probe because your mother does so. (Similarly, a lot of social interaction is not social learning.) If we equate culture to social learning, we have reduced culture to the least common denominator. Merely changing the label of social learning from *culture* to *tradition* does not change the argument (Fragaszy, 2003). Echoing Kroeber (1928), show me a school of guppies with something like innovation, dissemination, standardisation, durability, diffusion, and tradition, then the species will be taken seriously as a candidate for the Culture Club. The lesson learned is a variant on the old saw: something that explains almost everything explains nothing.

(17) Tradition and culture are a mess

As discussed above, some analysts use *tradition* and *culture* interchangeably. This is doubly wrong, if operationally one means by tradition learned behaviour that shows continuity across generations. There is tradition without culture, and culture without tradition.

Tradition without culture could be a migration route used by countless generations of chimpanzees, e.g. to the only waterhole left by the end of the dry season (cf. Mettke-Hofmann & Gwinner, 2003). A youngster could learn the optimal route from following others or from individual trial and error. Either explanation – cultural or noncultural – will explain the locomotor data, and just because the young ape is in the company of others does not mean that any knowledge is learned socially. (Any university lecturer can attest to this truism.)

Culture without tradition is seen easily by turning on the television, with its 'here today, gone-tomorrow' content. Popular culture thrives on fast horizontal transmission, especially among peers, and modern telecommunications allows such ideas (memes?) to go global. Just as rapidly, these cultural starbursts disappear. Does anyone remember pet rocks?

Some researchers have tried to set shorter timelines on traditions, in terms of months or years, e.g. Perry *et al.* (2003b). This ignores the key to the idea, i.e. intergenerational interval, whether in days or decades. To redefine tradition as some arbitrary period recalls Alice, in Lewis Carroll's *Through the Looking Glass*, when Humpty Dumpty says scornfully that a word can mean anything he wants it to mean. It resembles the reconfiguration of another time-word, *classic*, which in American culture means that anything that survives minimal longevity can be called a classic, including an item on a fast-food menu, e.g. McBurger's Double-Decker of catfish and dogfish. The lesson is that tradition takes time, if it is to mean anything, but no amount of time, short or long, necessarily makes it culture.

(18) *Language may be a red herring, or worse*

Language and culture go together like carcass and vulture. It is hard to imagine one without the other, for the simple reason that all known human groups have both. So, there is a perfect correlation between language and culture with no variation. There is an equally consistent relationship among bipedality and language and culture: all humans normally stand upright, speak and are cultural. Ditto for hairlessness, pentadactyly, etc. Coincidence tells us nothing causal, and until we find a human group with culture but lacking language, or vice versa, we can say little more about the relationship between the two for our species. But for apes, it is clear: culture exists without language.

In this context of cultural primatology and cultural anthropology, language is important in two ways, which sometimes are confused. First, for humans, the importance of language as the handmaiden of culture is undeniable. You cannot have oral history without speech. Using language, dialect, or inflection to distinguish between us and them is a basic human social strategy that can be of life or death importance. The fact that every one of us lives a daily life as a member of a human culture is unimaginable without language. Everything else is optional – we can be vegetarian or carnivore, nocturnal or diurnal, walk on our feet or hands – but we must have language. In contrast, however impressive their vocal and acoustic communication, nonhumans do not have language, in the sense of semantic and syntactic communication. But despite this, it seems that some other species do have culture, so language is a red herring.

Second, that language is important to cultural anthropology has been discussed above and will be only rehashed here. Most data in

sociocultural anthropology rely on verbal report: we listen to informants share their feelings and thoughts and try to make sense of them.
Methodologically, it is far easier to ask someone why he does something than it is to puzzle it out; it is far easier to ask someone what
happened yesterday at the annual fête than to wait until next year to
see what happens.

The problem is that people often do not tell the truth. Words are
just expirations with no inherent validity. Truth may come in shades
of falsehood, via omission, commission, or imprecision. Given this
problem, one might think that sociocultural anthropologists might try
to corroborate what people say by seeing what they do. This methodology of validating verbal report seems absent, with the conspicuous
exception of the behavioural ecologists. None of this is a problem for
the primatologist. So, what is advantaged: cultural anthropology with
discourse, or cultural primatology with observational data? If your life
depended on it, which would you trust: words or deeds? The lesson
is that language as culture is wonderful, but it still comes down to
inference.

(19) Teaching is the last resort, or even a poisoned chalice
Everyone loves the idea of teaching, especially educators. Critics of
cultural primatology (e.g. Premack (Premack & Premack, 1994), and
Tomasello (1999)) play tuition as a trump card when all else fails. It
has come to be the last bastion of uniquely human processes of information transmission, now that imitation has fallen (Byrne & Russon,
1998; Whiten *et al.*, 2003a). Critics point out rightly that the evidence for teaching in nonhumans is sparse, being at most anecdotal
(Boesch, 1991c; Guinet, 1991; Guinet & Bouvier, 1995; Maestripieri,
1996; Whiten, 1999). Unfortunately, the term *teaching* often is used
loosely. Here, I mean it to be when a knowledgeable individual (tutor)
acts so as to improve the understanding and performance of a less
knowledgeable individual (pupil). For a tighter operational definition,
see Caro and Hauser (1992). The embedded issues are: the tutor's
intentions, and the pupil's understanding; both are thorny. The tutor's

(intentional) modification of behaviour for the benefit of the pupil is altruistic; it is costly of time, energy, or risk. The pupil's comprehension implies not just more efficient performance, but also knowledge enhancement. (The latter is the difference between training and teaching. Using operant techniques, we can train chimpanzees to water-ski, but this is hardly teaching.)

Thus, it is easier to train than to teach; and it is even easier to lead by example. The simplest way to impart knowledge is to leave the task to the observational learner, as do most traditional human societies (Hewlett & Cavalli-Sforza, 1986). Only complicated, exacting and arbitrary sequential activities (e.g. rituals) require teaching through designated roles as in rites of passage. Ultimately, reliance on teaching is the price humans pay for literacy, i.e. externally stored symbolic knowledge. If we are to look for teaching in chimpanzees, it should be in customs, not in foraging or self-maintenance. The lesson for a large-brained creature – human or nonhuman – is not to yearn for teaching like some holy grail, but instead to avoid it unless absolutely necessary.

(20) Culture is a curse as well as a blessing

Anyone can recite the advantages of culture, and because it is Lamarckian rather than Darwinian, beneficial novelties can be added quickly, just as harmful ones can be dropped. Unlike genetic evolution that is tied to germ-cell lines, cultural evolution can spread in any direction. Culture is constrained only by learning capacity and companionship, while genes depend on the happy improbability of gamete fusion plus successful ontogeny. Most importantly, creatures with large-enough brains can be creative, while genes must wait for random mutations or other mimetic accidents to create novelty. All in all, culture sounds great.

The flipside to culture is sobering, even depressing. A collective of self-conscious creatures ensures the emergence of the 'tragedy of the commons'. Natural selection will favour individuals who take more than their share of communal resources. The inevitable strategy is: 'Get yours first, for it may not be there tomorrow.' Furthermore, with a hypertrophied brain comes self-deception, and such a clever mind can rationalise any action. The result is short-term gain, but at a long-term cost. Why do we follow suit? Because a key part of culture is conformity, which is the price we pay for the benefits of the community (Logan & Qirko, 1996).

Pestle-pounding of oil palms by the chimpanzees of Bossou (Yamakoshi & Sugiyama, 1995) looks at first like a case of cultural maladaptation, as discussed above, as the crown is mangled after being harvested. So, also, may the killing (rather than the punishing) of reproductive-age females from neighbouring communities, at least when done by males (Goodall et al., 1979). Maladaptation really has not yet been tackled in apes, as the compelling null hypothesis is always one of adaptation. The lesson is that the dubious distinction 'play now, pay later' may apply to cultural creatures other than humans, since cultural change may outstrip natural selection.

Perhaps the real example of cultural maladaptation in wild chimpanzees is crop-raiding. When humans clear chimpanzee habitats and plant edible cultigens in their place, it is not surprising that the apes exploit them. In many cases, the domesticated plants (e.g. bananas) are inserted piecemeal into existing forest, so they are hardly discernable as being artificial and 'off limits'. (Very few fields in African horticulture resemble the geometric landscapes of Western agribusiness; most are small and only partly cleared.) Crop-raiding only becomes maladaptive when humans defend their crops, and, when firearms are involved, the results of taking forbidden fruit can be fatal. No one seems to have studied systematically chimpanzee crop-raiding, although there is a literature on it for other species of primates (Maples et al., 1976a).

None of the twenty lessons presented in this chapter would have surfaced without a substantial body of knowledge now accrued in cultural primatology. At least, this is true for the well-studied chimpanzee. It is worth remembering that a single sentence here may encapsulate a decade of slowly accumulated, hard-won data. Every analytical comparison involves at least two datasets, and so doubles the workload. Further, the first sentence of the chapter needs reiterating: I have offered *opinions*, not conclusions. These are only interpretations, often based on sketchy data that need to be reconfigured as hypotheses and then taken back to the field for more testing. Finally, it would be misleading to finish this book by looking back on past accomplishments, however laudable. The next chapter takes a tentative look forward, to what may lie ahead for cultural primatology.

10 Does cultural primatology have a future?

There is plenty to do in cultural primatology, especially with chimpanzees. There are so many fragmented populations scattered across equatorial Africa from Senegal to Sudan, but so few are well known. Still, the number of cultures now described and somewhat studied is enough to think of doing multivariate analyses on the ethnographic data. For example, if termite-fishing and nut-cracking really are complementary in distribution, can this be explained by rainfall, latitude, etc? (It seems unlikely to be availability of prey: there appears to be an equivalent density of mounds of *Macrotermes* sp. at Gombe, where termite-fishing is customary, and Bossou, where it has been seen only once in 27 years!) How many points of origin are there for the widespread patterns of material culture, e.g. ant-dipping? We do not have the historical (or prehistorical?) data on actual origin of the customs anywhere, but this might be inferred from the geographical patterning of its distribution.

Co-ordination between field and laboratory – following the precedent set by Tetsuro Matsuzawa – should be expanded and extended. Why not set up a laboratory context in which it is advantageous for one ape to teach another, to see if teaching can play a major role in cultural transmission? Why not provide an over-supply of introduced hammer and anvil stones at wild nut-cracking sites, to see true preferences for size, shape, and weight? Why not let archaeologists excavate and map the lithic work sites of chimpanzees, then ask naïve judges to distinguish the fractured splinters of hammers and anvils from the flakes of cobbles and cores made by early hominids? Why not give zoo chimpanzees control over the lighting and air-conditioning of their indoor quarters to see if they would converge across colonies on a species-typical norm, or if different cultures of climate would emerge?

Future studies in cultural primatology should be methodologically ambitious and opportunistic. Reconsider the problem raised in earlier chapters: that we cannot do experiments in nature and so never will be able to draw firm conclusions about our subjects' observational learning. This seems intractable, but there are ways to tackle the problem. Suppose we looked carefully at the details of a given technique, e.g. pestle-pounding. That is, we use ethological scrutiny to discern the motor patterns, tempos, grips, sequences of components, iterations, etc. of how the palm frond is prepared, extracted, modified, employed, discarded, etc. Suppose we find individual differences in style across adult females, such that there is recognisably greater variation across subjects than within subjects. If there is observational learning going on, we should expect each offspring's style to be more like her or his mother than like any other adult female's. If there is only individual learning, perhaps prompted by stimulus enhancement, then there will be no such correlation. This approach is more than speculation: Byrne and his colleagues have used it repeatedly, producing elegant flowcharts to illustrate the sequences (Byrne & Byrne, 1993; Corp & Byrne, 2002).

More contact between cultural primatologists and palaeoanthropologists would help the latter in making better inferences about extinct hominoids and hominids. Dental microwear studies on nut-cracking versus non-nut-cracking chimpanzees might help draw inferences about the presence or absence of percussive technology in prehistory, since so often all that we have in the fossil record are teeth.

One would expect nut-cracking chimpanzees to have less damaged teeth. Sometimes even the previously ignored details can turn up contrasts, e.g. populations may exploit the same food items but use different techniques to do so. Fongoli's apes make huge wadges of hundreds of pellet-sized Allophyllus fruits, while Bossou's apes either do not wadge them at all, or produce wadges of only a few-score fruits.

Comparing the phytoliths and pollen in palaeosols with soil samples taken from various sites of chimpanzee field study, from savanna to rainforest, might help in palaeoecological reconstruction. Comparing the nut-cracking sites of humans and chimpanzees in the same forest might yield evidence of distinctive, diagnostic signatures in the artefacts. Comparing the subsistence technology of chimpanzees in areas in which they are the only ape species present, versus areas where they co-occur with gorillas, may be instructive about sites where more than one hominid species were sympatric. Similar comparative analysis could be done on populations of chimpanzees that are, and are not, sympatric with baboons. Similarly, studies of the means by which humans and apes today harvest the same forest resource (e.g. a major wild-fruiting crop (baobab, custard apple, hog plum)) may shed light on past interspecies competition. Just watching carefully how hammer stones and anvils are modified accidentally in the course of chimpanzees' nut-cracking, then collecting and refitting the debris, could be informative in helping to recognise the earliest stone artefacts. We may never have a time machine, but we share the planet with creatures whose links to our ancestors are closer than their links to us living *Homo sapiens*. Why not make more use of this?

We should make more use of the tools of modern molecular genetics now available to the field worker. Now that we can collect noninvasively the biological materials from free-ranging apes (hair, faeces, urine, spat-out wadges, saliva, leaves) and use the polymerase chain reaction (PCR) to amplify the DNA, we can genotype the individuals in a community, even if we never see them (McGrew *et al.*, 2004). Thus, the microsatellite data become a dependent variable providing information on kinship, group and party size, and sex ratio. For party composition, social network, and ranging, genotype becomes another independent variable, acting as a proxy for seeing the individual with others or at particular places. Finally, we can even use genotype to infer life-history, e.g. Yo is the only adult female who does not pestle-pound. On this basis, one might predict that she is an immigrant, who did not learn the pattern from growing up in the group. By coincidence, her son Yolo is the only grown-up male who does not pestle-pound, as he lacked a maternal model.

From what I can see, most if not all field primatologists are imprinted strongly upon the first population of chimpanzees that they came to know well. For me, as with so many other Western chimpologists, the touchstone is Gombe, or, more specifically, the apes of the Kasakela community. Although I never intended it to be that way, it seems that every new population of wild chimpanzees that I meet is compared automatically and contrasted with Fifi, Goblin, Freud, etc. This is only a problem if such perceptions became stereotypes, and as such, they constrain our thinking. From the chimpanzees of K-group, whom I first met in January 1975, to the chimpanzees of Bossou, whom I last met in August 2003, and all of those in between, I have learned much. Cultural primatology will benefit greatly from widespread and systematic exchanges of scientists and field assistants across study sites. Just as with *H. sapiens*, a person who knows only one culture of chimpanzees will have a limited view of the species.

Whether or not a useful dialogue between the cultural primatologist (CP) and the cultural anthropologist (CA) can be devised remains to be seen. At the very least, the CAs might tell the CPs what (if anything) the CPs could bring back from the forest that would interest the CAs. Consider the following hypothetical example: if after birth, the placenta was consumed by the new ape mother, neither CP or CA, would be surprised, as it is a common mammalian pattern. However, if the placenta was shared with the sister of the mother, at least the CP would be interested, for this is not the usual pattern of meat-sharing. The CP probably would publish the tentative report and seek comparative data from other populations. If, instead, the placenta were handed over to the oldest female in the ape group, regardless of whether or not she shared kinship ties with the mother, the CA might be forced to join the CP in paying attention, for there is no biological reason for such an arbitrary act. Finally, if the placenta was not eaten, but instead buried with the community in attendance, and all apes present in turn stamped on the burial spot, one wonders which – the CA or CP – would take the senior authorship of the report to follow. The whole range of possibilities just presented is plausible; the question is: how far along the continuum would one go before choosing to publish the results in an anthropology, rather than primatology, journal?

What seems to be emerging is a three-stage process, two stages of which are recognisable. Until the 1970s, we thought of *Pan troglodytes* as a species of great ape about which there was much to learn. In traditional zoological fashion, we sought information about the chimpanzee. Then we discovered diversity – haphazardly at first – but by the 1990s systematically. Instead of The Chimpanzee, there were many chimpanzees – individuals, groups, populations, etc. Moreover, the race was on to report new behavioural patterns, and then to throw them into the public domain to see if anyone else had them, too, in their study population of apes. This emphasis on diversity changed from cataloguing and classifying and coding to hypothesis-testing at the turn of the twenty-first century. This species-wide analysis of diversity will increase in power, once the data are available widely (on the web), and secondary scientists (not just the field workers) realise the treasure trove that is now available. From this will emerge the next stage, in which commonality comes out of diversity. From the features that turn up again and again, in eastern, central, and western Africa will come finally a knowledge of chimpanzee nature, locally nuanced but essentially the same: the essence of our closest living relation.

CULTURAL SURVIVAL

All of the above will be academic, however, if the chimpanzee becomes extinct. All the curiosity and ambition in the world in cultural primatology will be powerless, except as a retrospective exercise, unless we increase our efforts to preserve apes. If we do no more than we are doing now, it will not be enough, and the chimpanzee will go down the drain, as finally as the dodo (Wrangham *et al.*, 2003). Palaeoanthropologists who debate the origins of fire in human evolution are forced to rely upon what they can glean from the palaeontological and archaeological records. They have no choice, as their subjects are long gone, but we primatologists do have a choice, thankfully.

Traditional conservation based on the value of biodiversity or the wonders of nature is not enough, any more than are ecological arguments based on apes as predators, prey, seed dispersers, etc. Arguments based on the apes' intelligence, or their entertaining acts, or their use as biomedical models, or their earning power for ecotourism, are not enough. We must protect them because their cultural diversity and complexity are all that we have today to help us understand the origins and development of our own cultural evolution. We can infer soft tissues from hard, or elementary technology from artefacts, but palaeobehaviour, especially sociality, is invisible.

So, what must we do? We must preserve, not just conserve. We must protect and maintain whole ecosystems, full of predators, prey, and competitors, that reflect the range of the chimpanzee as a species. We must keep safe cultures as well as populations, and when possible preserve populations big enough for intergroup gene and meme flow. Where populations have fragmented, we must restore corridors to reconnect them; where communities are threatened we must erect barriers to keep out hazards. Where people have encroached into nature reserves, they must be displaced. Where tourism presents a threat to the health of apes, it must be cleaned up or stopped. And who is to pay for this, you ask? All of us, but especially the scientists who have been so lucky to be able to study these wonderful creatures. Accordingly, all the profits of this book will go to help chimpanzees in Africa. This is not meant to sound misanthropic, but the fact is that the number of *H. sapiens* on planet Earth exceeds 6 billion souls and is rising. The number of chimpanzees is probably less than 100 000 and is falling (Walsh *et al.*, 2003). Who deserves the help and attention?

I ended a previous book (McGrew, 1992, p. 230) with the following sentences: 'Our hominid predecessors are irretrievably gone, while our hominoid cousins (just) survive. What a pity it would be to extinguish them before they could tell us all that they know.' This statement seems more apt than ever. I end this book with a plea: if even one-tenth of what you read in this book impresses you, then do something about it. If you don't know what to do, then let me know and I will make a suggestion or two.

References

Alp, R. (1993). Meat eating and ant dipping by wild chimpanzees in Sierra Leone. *Primates*, **34**, 463–8.

(1997). 'Stepping-sticks' and 'seat-sticks': new types of tools used by wild chimpanzees (*Pan troglodytes*) in Sierra Leone. *American Journal of Primatology*, **41**, 45–52.

(1999). Update on Tacugama chimpanzee sanctuary. International Primate Protection League *Newsletter*, **26**, 19.

Alvard, M. S. (2003). The adaptive nature of culture. *Evolutionary Anthropology*, **12**, 136–49.

Anderson, J. R., Williamson, E. A. & Carter, J. (1983). Chimpanzees of Sapo Forest, Liberia: density, nests, tools and meat-eating. *Primates*, **24**, 594–601.

Arcadi, A. C. (1996). Phrase structure of wild chimpanzee pant hoots: patterns of production and interpopulation variability. *American Journal of Primatology*, **39**, 159–78.

(2000). Vocal responsiveness in male wild chimpanzees: implications for the evolution of language. *Journal of Human Evolution*, **39**, 205–33.

Arcadi, A. C. & Mugurusi, F. (2004). Acoustic differences between buttress drumming and ground slapping by an adult male chimpanzee. *Primates, forthcoming*.

Arcadi, A. C., Robert, D. & Mugurusi, F. (2004). A comparison of buttress drumming by male chimpanzees from two populations. *Primates, forthcoming*.

Arcadi, A. C. & Wrangham, R. W. (1999). Infanticide in chimpanzees: review of cases and a new within-group observation from the Kanyawara study group in Kibale National Park. *Primates*, **40**, 337–51.

Asquith, P. A. (1989). Provisioning and the study of free-ranging primates: history, effects, and prospects. *Yearbook of Physical Anthropology*, **32**, 129–58.

Assersohn, C., Whiten, A., Kiwede, Z. T., Tinka, J. & Karamagi, J. (2004). Use of leaves to inspect ectoparasites in wild chimpanzees: a third cultural variant? *Primates, forthcoming*.

Aunger, R. (1994). Are food avoidances maladaptive in the Ituri Forest of Zaire? *Journal of Anthropological Research*, **50**, 277–310.

(1995). On ethnography: story-telling or science? *Current Anthropology*, **36**, 97–130.

Azuma, S. & Toyoshima, A. (1961–2). Progress report on the survey of chimpanzees in their natural environment, Kabogo Point area, Tanganyika. *Primates*, **3**, 61–70.

Backwell, L. R. & d'Errico, F. (2001). Evidence of termite foraging by Swartskrans early hominids. *Proceedings of the National Academy of Sciences (USA)*, **98**, 1358–63.

Baird, R. W. (2000). The killer whale: foraging specializations and group hunting. In: *Cetacean Societies: Field Studies of Dolphins and Whales*, eds. J. Mann, R. C. Connor, P. L. Tyack, & H. Whitehead, Chicago: University of Chicago Press, pp. 127–53.

Baldwin, P. J. (1979). The natural history of the chimpanzee (*Pan troglodytes verus*) at Mt. Assirik, Senegal. Ph.D. Thesis, University of Stirling. (*Dissertation Abstracts International*, **c44**, 92, 1983.)

Baldwin, P. J., Sabater Pi, J., McGrew, W. C. & Tutin, C. E. G. (1981). Comparisons of nests made by different populations of chimpanzees (*Pan troglodytes*). *Primates*, **22**, 474–86.

Bauer, H. R. (1977). Chimpanzee bipedal locomotion in the Gombe National Park, East Africa. *Primates*, **18**, 913–21.

Beatty, H. (1951). A note on the behavior of the chimpanzee. *Journal of Mammalogy*, **32**, 118.

Beck, B. B. (1980). *Animal Tool Behavior*. New York: Garland STPM Press.

Benedict, R. (1935). *Patterns of Culture*. London: Routledge & Kegan Paul.

van Bergen, Y., Laland, K. N. & Hoppitt, W. (2004). Social learning, innovation and intelligence in fish. In: *Comparative Vertebrate Cognition: Are Primates Superior to Non-Primates?*, eds. L. J. Rogers & G. Kaplan, *forthcoming*.

Berlin, B. & Kay P. (1969). *Basic Colour Terms: Their Universality and Evolution*. Berkeley, CA: University of California Press.

Bermejo, M. & Illera, G. (1999). Tool-set for termite-fishing and honey extraction by wild chimpanzees in the Lossi Forest, Congo. *Primates*, **40**, 619–27.

Bermejo, M., Illera, G. & Sabater Pi, J. (1989). New observations on the tool-behavior of the chimpanzees from Mt. Assirik (Senegal, West Africa). *Primates*, **30**, 65–73.

Bernstein, I. S. (1967). Age and experience in chimpanzee nest building. *Psychological Reports*, **20**, 1106.

(1969). A comparison of nesting patterns among the great apes. In: *The Chimpanzee*, Vol. 1: *Anatomy, Behavior and Diseases* of *Chimpanzees*, ed. G. H. Bourne, Basel: S. Karger, pp. 393–402.

Biro, D. & Matsuzawa, T. (2001). Chimpanzee numerical competence: cardinal and ordinal skills. In: *Primate Origins of Human Cognition and Behavior*, ed. T. Matsuzawa, Tokyo: Springer, pp. 199–225.

Blackmore, S. (1999). *The Meme Machine*. Oxford: Oxford University Press.

Blumenschine, R. J. & Selvaggio, M. M. (1988). Percussion marks on bone surfaces as a diagnostic of hominid behaviour. *Nature*, **333**, 763–5.

Boesch, C. (1978). Nouvelles observations sur les chimpanzés de la forêt de Taï (Côte d'Ivoire). *La terre et la vie*, **32**, 195–201.

(1991a). Handedness in wild chimpanzees. *International Journal of Primatology*, **12**, 541–80.

(1991b). Symbolic communication in wild chimpanzees? *Human Evolution*, **6**, 81–90.

(1991c). Teaching among wild chimpanzees. *Animal Behaviour*, **41**, 530–2.

(1996). Three approaches for assessing chimpanzee culture. In: *Reaching into Thought: The Minds of Great Apes*, eds. A. E. Russon, K. A. Bard & S. T. Parker, Cambridge: Cambridge University Press, pp. 404–29.

(2002). Cooperative hunting roles among Taï chimpanzees. *Human Nature*, **13**, 27–46.

(2003). Is culture a golden barrier between human and chimpanzee? *Evolutionary Anthropology*, **12**, 82–91.

Boesch, C. & Boesch, H. (1983). Optimisation of nut-cracking with natural hammers by wild chimpanzees. *Behaviour*, **83**, 265–86.

(1984a). Mental map in wild chimpanzees: an analysis of hammer transports for nut cracking. *Primates*, **25**, 160–70.

(1984b). Possible causes of sex differences in the use of natural hammers by wild chimpanzees. *Journal of Human Evolution*, **13**, 415–40.

(1990). Tool use and tool making in wild chimpanzees. *Folia Primatologica*, **54**, 86–99.

(1993). Different hand postures for pounding nuts with natural hammers by wild chimpanzees. In: *Hands of Primates*, eds. H. Prueschoft & D. J. Chivers, New York: Springer, 31–43.

Boesch, C. & Boesch-Achermann, H. (2000). *The Chimpanzees of the Tai Forest. Behavioral Ecology and Evolution*. Oxford: Oxford University Press.

Boesch, C., Marchesi, P., Marchesi, N., Fruth, B. & Joulian, F. (1994). Is nut-cracking in wild chimpanzees a cultural behaviour? *Journal of Human Evolution*, **26**, 325–38.

Boinski, S. (1988). Use of a club by a wild white-faced capuchin (*Cebus capucinus*) to attack a venomous snake (*Bothrops asper*). *American Journal of Primatology*, **14**, 177–9.

Boinski, S., Quatrone, R. P. & Swartz, H. (2000). Substrate and tool use by brown capuchins in Suriname: ecological contexts and cognitive bases. *American Anthropologist*, **102**, 741–61.

Bonner, J. T. (1980). *The Evolution of Culture in Animals*. Princeton, NJ: Princeton University Press.

Borner, M. (1985). The rehabilitated chimpanzees of Rubondo Island. *Oryx*, **19**, 151–4.

Borofsky, R., Barth, R., Shweder, R. A., Rodseth, L. & Stolgenberg, N. M. (2001). WHEN: a conversation about culture. *American Anthropologist*, **103**, 432–46.

Boyd, R. & Richerson, P. J. (1985). *Culture and the Evolutionary Process*. Chicago: University of Chicago Press.

(1996). Why culture is common, but cultural evolution is rare. *Proceedings of the British Academy*, **88**, 77–93.

Boysen, S. T. & Hallberg, K. I. (2000). Primate numerical competence: contributions toward understanding nonhuman cognition. *Cognitive Science*, **24**, 423–43.

Brewer, S. M. (1978). *The Chimps of Mt. Asserik*. New York: Alfred A. Knopf.

(1982). Essai de réhabilitation au Parc National du Niokolo-Koba des chimpanzés auparavant en captivité. *Memoires de l'Institut Fondamental d'Afrique Noire*, **92**, 341–62.

Brewer, S. M. & McGrew, W. C. (1990). Chimpanzee use of a tool-set to get honey. *Folia Primatologica*, **54**, 100–4.

Brown, D. E. (1991). *Human Universals*. New York: McGraw-Hill.

Brown, G. & Laland, K. N. (2003). Social learning in fishes: a review. *Fish and Fisheries*, **4** (3), 280–8.

Bshary, R., Wickler, W. & Fricke, H. (2002). Fish cognition: a primate eye's view. *Animal Cognition*, **5**, 1–13.

Bunn, H. T. (1981). Archaeological evidence for meat-eating by Plio-Pleistocene hominids from Koobi Fora and Olduvai Gorge. *Nature*, **291**, 574–7.

Burrell, K. (1998). Cultural variation in savannah sparrow, *Passerculus sandwichensis*, songs: an analysis using the meme concept. *Animal Behaviour*, **56**, 995–1003.

Busse, C. D. (1978). Do chimpanzees hunt cooperatively? *American Naturalist*, **112**, 767–70.

Bygott, J. D. (1972). Cannibalism among wild chimpanzees. *Nature*, **238**, 410–11.

Byrne, R. W. (1995). *The Thinking Ape: Evolutionary Origins of Intelligence*. Oxford: Oxford University Press.

Byrne, R. W. & Byrne, J. M. E. (1993). Complex leaf-gathering skills of mountain gorillas (*Gorilla g. beringei*): variability and standardization. *American Journal of Primatology*, **31**, 241–61.

Byrne, R. W. & Russon, A. E. (1998). Learning by imitation: a hierarchical approach. *Behavioral and Brain Sciences*, **21**, 667–721.

Byrne, R. W. & Whiten, A. (1988). *Machiavellian Intelligence. Social Expertise and the Evolution of Intellect in Monkeys, Apes, and Humans*. Oxford: Oxford University Press.

Caldwell, C. A. & Whiten, A. (2002). Evolutionary perspectives on imitation: is a comparative psychology of social learning possible? *Animal Cognition*, **5**, 193–208.

Caro, T. M. & Hauser, M. D. (1992). Is there teaching in nonhuman animals? *Quarterly Review of Biology*, **67**, 151–74.

Chagnon, N. A. (1992). *Yanomamö*. 4th edn. Fort Worth, TX: Harcourt Brace Jovanovich.

Chevalier-Skolnikoff, S. (1990). Tool use by wild *Cebus* monkeys at Santa Rosa National Park, Costa Rica. *Primates*, **31**, 375–83.

Clark, J. D., Beyene, Y., WoldeGabriel, G., Hart, W. K., Renne, P. R., Gilbert, H., Defluer, A., Suwa, G., Katoh, S., Ludwig, K. R., Boisserie, J-R., Asfaw, B. & White, T. D. (2003). Stratigraphic, chronological and behavioural contexts of Pleistocene *Homo sapiens* from Middle Awash, Ethiopia. *Nature*, **423**, 747–52.

Clarke, M. R., Daigle, R. M. & McGrew, W. C. (1993). Stone-handling and cheekpouch filling in a colony of rhesus monkeys. *American Journal of Physical Anthropology*. Suppl. 16, 71, (abstract.)

Conner, R. C. & Smolker, R. A. (1985). Habituated dolphins (*Tursiops* sp.) in Western Australia. *Journal of Mammology*, **6**, 398–400.

Connor, R. C., Smolker, R. A. & Richards, A. F. (1992). Two levels of alliance formation among male bottlenose dolphins (*Tursiops* sp.) *Proceedings of the National Academy of Sciences (USA)*, **89**, 987–90.

Connor, R. C., Wells, R. S., Mann, J. & Read, A. J. (2000). The bottlenose dolphin: social relationships in a fission–fusion society. In: *Cetacean Societies: Field Studies of Dolphins and Whales*, eds. J. Mann, R. C. Connor, P. L. Tyack, & H. Whitehead, Chicago: University of Chicago Press, pp. 91–126.

Corp, N. & Byrne, R. W. (2002). The ontogeny of manual feeding skill in wild chimpanzees: evidence from feeding on the fruit of *Saba florida*. *Behaviour*, **139**, 137–68.

Cronk, L. (1999). *That Complex Whole. Culture and the Evolution of Human Behavior.* Boulder, CO.: Westview Press.

Dawkins, R. (1976). *The Selfish Gene*. Oxford: Oxford University Press.

Deecke, V. B., Slater, P. J. B. & Ford, J. K. B. (2002). Selective habituation shapes acoustic predator recognition in harbour seals. *Nature*, **420**, 171–3.

Dennett, D. C. (1996). *Darwin's Dangerous Idea. Evolution and the Meanings of Life.* New York: Simon & Schuster.

Devos, C., Gatti, S. & Levrero, F. (2002). New record of algae feeding and scooping by *Pan t. trogolodytes* at Lokoue Bai in Odzala National Park, Republic of Congo. *Pan Africa News*, **9**, 19–21.

Diamond, J. (1986). Animal art: variation in bower style among male bowerbirds *Amblyornis inornatus*. *Proceedings of the National Academy of Sciences (USA)*, **83**, 3042–6.

(1988). Experimental study of bower decoration by the bowerbird *Amblyornis inornatus*, using colored poker chips. *American Naturalist*, **131**, 631–53.

Diezinger, F. & Anderson, J. R. (1986). Starting from scratch: a first look at a 'displacement activity' in group-living rhesus monkeys. *American Journal of Primatology*, **11**, 117–24.

Dunbar, R. I. M. (1992). Functional significance of social grooming in primates. *Folia Primatologica*, **57**, 121–31.

(1996). *Grooming, Gossip and the Origin of Language*. London: Faber & Faber.

Dunbar, R. I. M. & Sharman, M. J. (1984). Is social grooming altruistic? *Zeitschrift für Tierpsychologie*, **64**, 163–73.

Durham, W. (1991). *Coevolution. Genes, Culture, and Human Diversity*. Stanford, CA: Stanford University Press.

Eibl-Eibesfeldt, I. (1989). *Human Ethology*. New York: Aldine de Gruyter.

Emery, N. J. & Clayton, N. S. (2001). Effects of experience and social context on prospective caching strategies by scrub jays. *Nature*, **414**, 443–6.

Fay, J. M. & Carroll, R. W. (1994). Chimpanzee tool use for honey and termite extraction in central Africa. *American Journal of Primatology*, **34**, 309–17.

Fernandes, M. E. B. (1991). Tool use and predation of oysters (*Crassostrea rhizophorea*) by the tufted capuchin, *Cebus apella apella*, in brackish water mangrove swamp. *Primates*, **32**, 529–31.

Fiorito, G. & Scotto, P. (1992). Observational learning in *Octopus vulgaris*. *Science*, **256**, 545–7.

Fisher, J. & Hinde, R. A. (1949). The opening of milk bottles by birds. *British Birds*, **42**, 347–57.

Foley, R. A. (1987). *Another Unique Species. Patterns in Human Evolutionary Ecology*. Harlow: Longman.

(2001). Evolutionary perspectives on the origins of human social institutions. *Proceedings of the British Academy*, **110**, 171–95.

(2002). Adaptive radiations and dispersals in hominin evolutionary ecology. *Evolutionary Anthropology*, Suppl. 1, 32–7.

(2003). The emergence of culture in the context of hominin evolutionary patterns. In: *Evolution and Culture*, ed. S. Levinson & P. Jaisson. Cambridge, MA: MIT Press, pp. 25–42.

Foley, R. A. & Lahr, M. M. (2003). On stony ground: lithic technology, human evolution, and the emergence of culture. *Evolutionary Anthropology*, **12**, 109–22.

Fossey, D. (1983). *Gorillas in the Mist*. Boston: Houghton Mifflin.

Fragaszy, D. (2003). Making space for traditions. *Evolutionary Anthropology*, **12**, 61–70.

Fragaszy, D. & Perry, S. (2003). Towards a biology of traditions. In: *The Biology of Traditions: Models and Evidence*, eds. D. M. Fragaszy, & S. Perry, Cambridge: Cambridge University Press, pp. 1–32.

Fragaszy, D. M., Visalberghi, E. & Fedigan, L. M. (eds.) (2004). *The Complete Capuchin: The Biology of the Genus Cebus*. Cambridge: Cambridge University Press.

Freeberg, T. M. (2000). Culture and courtship in vertebrates: a review of social learning and transmission of courtship systems and mating patterns. *Behavioral Processes*, **51**, 177–92.

Fruth, B. & Hohmann, B. (1994). Comparative analyses of nest building behavior in bonobos and chimpanzees. In: *Chimpanzee Cultures*, eds. R. W. Wrangham, W. C. McGrew, F. B. M. de Waal, & P. G. Heltne, Cambridge, MA: Harvard University Press, pp. 109–28.

(1996). Nest building behaviour in the great apes: the great leap forward? In: *Great Ape Societies*, eds. W. C. McGrew, L. F. Marchant, & T. Nishida, Cambridge: Cambridge University Press, pp. 225–40.

Fuentes, A. & Wolfe, L. D. (eds.) (2002). *Primates Face to Face: The Conservation Implications of Human–Nonhuman Primate Interconnections*. New York: Cambridge University Press.

Galdikas, B. M. F. (1982). Orangutan tool use at Tanjung Puting Reserve, Central Indonesian Borneo (Kalimantan Tengah). *Journal of Human Evolution*, **10**, 19–33.

(1989). Orangutan tool use. *Science*, **243**, 152.

Galef, B. G. (1992). The question of animal culture. *Human Nature*, **3**, 157–78.

(2001). Where's the beef? Evidence of culture, imitation, and teaching in cetaceans? *Behavioral and Brain Sciences*, **24**, 335.

(2003). Social learning in animals. In: *Encyclopedia of Cognitive Science*, ed. L. Nadel. London: Nature Publishing Group, pp. 74–80.

Gallup, G. G. (1970). Chimpanzees: self-recognition. *Science*, **167**, 86–7.

Gonder, M. K., Oates, J. F., Disotell, T. R., Forstner, M. R. J., Morales, J. C. & Melnick, D. J. (1997). A new West African chimpanzee subspecies? *Nature*, **388**, 337.

Goodall, J. (1962). Nest building behavior in the free ranging chimpanzee. *Annals of the New York Academy of Sciences*, **102**, 455–67.

(1963). Feeding behaviour of wild chimpanzees. *Symposia of the Zoological Society of London*, **10**, 39–48.

(1964). Tool-using and aimed throwing in a community of free-living chimpanzees. *Nature*, **201**, 1264–6.

(1967). *My Friends the Wild Chimpanzees*. Washington, DC: National Geographic Society.

(1968). The behaviour of free-living chimpanzees in the Gombe Stream Reserve. *Animal Behaviour Monographs*, **1**, 161–311.

(1971). *In the Shadow of Man*. London: William Collins.

(1973). Cultural elements in a chimpanzee community. In: *Precultural Primate Behavior*, ed. E. W. Menzel, Basel: S. Karger, pp. 144–84.

(1977). Infant killing and cannibalism in free-living chimpanzees. *Folia Primatologica*, **28**, 259–82.

(1982). Order without law. *Journal of Social and Biological Structures*, **5**, 353–60.

(1986a). Social rejection, exclusion, and shunning among the Gombe chimpanzees. *Ethology and Sociobiology*, **7**, 227–36.

(1986b). *The Chimpanzees of Gombe: Patterns of Behavior*. Cambridge, MA: Harvard University Press.

(2003). Fifi fights back. *National Geographic Magazine*, **203**(4), 76–89.

Goodall, J., Bandoro, A., Bergmann, E., Busse, C., Matama, H., Mpongo, E., Pierce, A., & Riss, D. (1979). Intercommunity interactions in the chimpanzee population of the Gombe National Park. In: *The Great Apes*, eds. D. A. Hamburg & E. R. McCown, Menlo Park, CA: Benjamin/Cummings, pp. 13–53.

Goren-Inbar, N., Sharon, G., Melamed, Y. & Kislev, M. (2002). Nuts, nut cracking, and pitted stones at Gesher Benot Ya'agov, Israel. *Proceedings of the National Academy of Sciences (USA)*, **99**, 2455–60.

Green, S. (1975). Dialects in Japanese monkeys: vocal learning and cultural transmission of locale-specific behavior? *Zeitschrift für Tierpsychologie*, **38**, 301–14.

Guenther, M. M. & Boesch, C. (1993). Energetic cost of nut-cracking behaviour in wild chimpanzees. In: *Hands of Primates*, eds. H. Prueschoft & D. J. Chivers, New York: Springer, pp. 109–29.

Guinet, C. (1991). Intentional stranding apprenticeship and social play in killer whales (*Orcinus orca*). *Canadian Journal of Zoology*, **69**, 2712–16.

Guinet, C. & Bouvier, J. (1995). Development of intentional stranding hunting techniques in killer whale (*Orcinus orca*) calves at Crozet Archipelago. *Canadian Journal of Zoology*, **73**, 27–33.

Hall, K. R. L. & Schaller, G. B. (1964). Tool-using behavior of the California sea otter. *Journal of Mammalogy*, **45**, 287–98.

Hamai, M., Nishida, T., Takasaki, H. & Turner, L. A. (1992). New records of within-group infanticide and cannibalism in wild chimpanzees. *Primates*, **33**, 151–62.

Hamilton, W. D. (1971). Geometry for the selfish herd. *Journal of Theoretical Biology*, **31**, 295–311.

Hannah, A. C. & McGrew, W. C. (1987). Chimpanzees using stones to crack open palm nuts in Liberia. *Primates*, **28**, 31–46.

(1989). Rehabilitation of captive chimpanzees. In: *Primate Responses to Environmental Change*, ed. H. O. Box, London: Chapman & Hall, pp. 167–86.

Hansell, M. H. (1984). *Animal Architecture and Building Behaviour*. London: Longman.

(2004). *Animal Architecture*. Oxford: Oxford University Press.

Harding, R. S. O. (1984). Primates of the Kilimi area, northwest Sierra Leone. *Folia Primatologica*, **42**, 96–114.

Harris, M. (1964). *The Nature of Cultural Things*. New York: Random House.

(1979). *Cultural Materialism*. New York: Vintage.

Hart, H. & Panzer, A. (1925). Have subhuman animals culture? *American Journal of Sociology*, **30**, 703–9.

Hashimoto, C., Furuichi, T. & Tashiro, Y. (2001). What factors affect the size of chimpanzee parties in the Kalinzu Forest, Uganda? Examination of fruit abundance and number of estrus females. *International Journal of Primatology*, **22**, 947–59.

Hawkes, K., O'Connell, J. F., Blurton Jones, N. G., Alvarez, H. & Charnov, E. L. (1998). Grandmothering, menopause, and the evolution of human life history strategies. *Proceedings of the National Academy of Sciences* (USA), **95**, 1336–9.

Hayes, K. J. & Hayes, C. (1954). The cultural capacity of chimpanzee. *Human Biology*, **26**, 288–303.

de Heinzelin, J., Clark, J. D., White, T., Hart, W., Renne, P., WoldeGabriel, G., Beyene, Y. & Vrba, E. (1999). Environment and behavior of 2.5 million-year old Bouri hominids. *Science*, **284**, 625–9.

Helfman, G. S. & Schultz, E. T. (1984), Social tradition of behavioural traditions in coral reef fish. *Animal Behaviour*, **32**, 379–84.

Hewlett, B. B. & Cavalli-Sforza, L. L. (1986). Cultural transmission among Aka pygmies. *American Anthropologist*, **88**, 922–34.

Hinde, R. A. (1987). *Individuals, Relationships and Culture: Links between Ethology and the Social Sciences*. Cambridge: Cambridge University Press.

Hirata, S., Myowa, M. & Matsuzawa, I. (1998). Use of leaves as cushions to sit on wet ground by wild chimpanzees. *American Journal of Primatology*, **44**, 215–20.

Hirata, S., Watanabe, K. & Kawai, M. (2001a). 'Sweet-potato washing' revisited. In: *Primate Origins of Human Cognition and Behavior*, ed. T. Matsuzawa, Tokyo: Springer, pp. 487–508.

Hirata, S., Yamakoshi, G., Fujita, S., Ohasi, G. & Matsuzawa, T. (2001b). Capturing and toying with hyraxes (*Dendrohyrax dorsalis*) by wild chimpanzees (*Pan troglodytes*) at Bossou, Guinea. *American Journal of Primatology*, **53**, 93–7.

Hladik, C. M. (1973). Alimentation et activité d'un groupe de chimpanzés reintroduits en forêt Gabonaise. *La terre et la vie*, **27**, 343–413.

Hladik, C. M. & Viroben, G. (1974). L'alimentation protéique du chimpanzé dans son environnement forestier natural. *Comptes Rendes, Serie D*, **279**, 1475–8.

Hockett, C. F. (1960). The origin of speech. *Scientific American*, **203**, 89–96.

(1973). *Man's Place in Nature*. New York: McGraw-Hill.

Hohmann, G. & Fruth, B. (2003). Culture in bonobos? Between-species and within-species variation in behavior. *Current Anthropology*, **44**, 563–71.

Huffman, M. A. (1984). Stone-play of *Macaca fuscata* in Arashiyama B troop: transmission of a non-adaptive behavior. *Journal of Human Evolution*, **13**, 725–35.

(1996). Acquisition of innovative cultural behaviors in nonhuman primates: a case study of stone handling, a socially transmitted behavior in Japanese macaques. In: *Social Learning in Animals: The Roots of Culture*, eds. B. G. Galef & C. M. Heyes, Orlando, FL: Academic Press, pp. 267–89.

(1997). Current evidence for self-medication in primates: a multidisciplinary perspective. *Yearbook of Physical Anthropology*, **40**, 171–200.

Huffman, M. A. & Hirata, S. (2003). Biological and ecological foundations of primate behavioral traditions. In: *The Biology of Traditions: Models and Evidence*, eds. D. M. Fragaszy, & S. Perry, Cambridge: Cambridge University Press, pp. 267–96.

Huffman, M. A. & Seifu, M. (1989). Observations on the illness and consumption of a medicinal plant *Vernonia amygdalina* by a wild chimpanzee in the Mahale Mountains, Tanzania. *Primates*, **30**, 51–63.

Humle, T. & Matsuzawa, T. (2001). Behavioral diversity among the wild chimpanzee populations of Bossou and neighbouring areas, Guinea and Cote d'Ivoire, West Africa. A preliminary report. *Folia Primatologica*, **72**, 57–68.

(2002). Ant-dipping among the chimpanzees of Bossou, Guinea, and comparisons with other sites. *American Journal of Primatology*, **58**, 133–48.

(2004). Oil palm use by adjacent communities of chimpanzees at Bossou and Nimba Mountains, West Africa. *International Journal of Primatology*, **25**, 551–81.

Humphries, S. & Ruxton, G. D. (1999) Bower-building: coevolution of display traits in response to the costs of female choice? *Ecology Letters*, **2**, 404–13.

Hunt, G. R. (1996). Manufacture and use of hook-tools by New Caledonian crows. *Nature*, **379**, 249–51.

Hunt, G. R. & Gray, R. D. (2003). Diversification and cumulative evolution in New Caledonian crow tool manufacture. *Proceedings of the Royal Society B*,

Hunt, K. D. (1992). Positional behavior of *Pan troglodytes* in the Mahale Mountains and Gombe Stream National Parks, Tanzania. *American Journal of Physical Anthropology*, **87**, 83–105.

(1994). The evolution of human bipedality: ecology and functional morphology. *Journal of Human Evolution*, **26**, 183–202.

Hunt, K. D. & McGrew, W. C. (2002). Chimpanzees in the dry areas of Assirik, Senegal and Semliki Wildlife Reserve, Uganda: In: *Behavioural Diversity in Chimpanzees and Bonobos*, eds. C. Boesch, G. Hohmann & L. F. Marchant, Cambridge: Cambridge University Press, pp. 35–51.

Ihobe, H. (1992). Observations on the meat-eating behavior of wild bonobos (*Pan paniscus*) at Wamba, Republic of Zaire. *Primates*, **33**, 247–50.

Ingmanson, E. J. (1996). Tool-using behavior in wild *Pan paniscus*: social and ecological considerations. In: *Reaching into Thought: The Minds of Great Apes*, eds. A. R. Russon, K. A. Bard, & S. T. Parker, Cambridge: Cambridge University Press, pp. 190–210.

Ingold, T. (1986). *The Appropriation of Nature. Essays on Human Ecology and Social Relations*. Manchester: Manchester University Press.

(2001). The use and abuse of ethnography. *Behavioral and Brain Sciences*, **24**, 337.

Inoue-Nakamura, N. (1997). Mirror self-recognition in nonhuman primates: a phylogenetic approach. *Japanese Psychological Research*, **39**, 266–75.

Isaac, G. L. (1978). The food-sharing behavior of protohominid hominids. *Scientific American*, **238** (4), 90–108.

Itani, J. (1979). Distribution and adaptation of chimpanzees in an arid area. In: *The Great Apes*, eds. D. A. Hamburg & E. R. McCown, Menlo Park, CA: Benjamin/Cummings, pp. 55–71.

Itani, J. & Nishimura, A. (1973). The study of infrahuman culture in Japan. In: *Precultural Primate Behavior*, ed. E. W. Menzel, Basel: S. Karger, pp. 26–50.

Itani, J. & Suzuki, A. (1967). The social unit of chimpanzees. *Primates*, **8**, 355–81.

Janik, V. M. (1999). Origins and implications of vocal learning in bottlenose dolphins. In: *Mammalian Social Learning: Comparative and Ecological Approaches*, ed. H. O. Box & K. R. Gibson, Cambridge: Cambridge University Press, pp. 308–26.

(2000). Whistle matching in wild bottlenose dolphins. *Science*, **289**, 1355–7.

Janik, V. M. & Slater, P. J. B. (1997). Vocal learning in mammals. *Advances in the Study of Behavior*, **26**, 59–99.

Jolly, A. (1972). Hour of birth in primates and man. *Folia Primatologica*, **18**, 108–21.

Jones, C. & Sabater Pi, J. (1971). Comparative ecology of *Gorilla gorilla* (Savage and Wyman) and *Pan troglodytes* (Blumenbach) in Rio Muni, West Africa. *Bibliotheca Primatologica*, **13**, 1–96.

Joulian, F. (1994). Culture and material culture in chimpanzees and early hominids. In: *Current Primatology*, Vol. II: *Social Development, Learning and Behaviour*, eds. J. J. Roeder, B. Thierry, J. R. Anderson & N. Herrenschmidt, Strasbourg: Université Louis Pasteur, pp. 397–404.

Joulian, F. (1996). Comparing chimpanzee and early hominid techniques: some contributions to cultural and cognitive questions. In: *Modelling the Early Human Mind*, eds. P. Mellars & K. Gibson, Cambridge: McDonald Institute Monographs, pp. 173–89.

Kaplan, H. & Hill, K. (1985). Hunting ability and reproductive success among male Ache foragers. *Current Anthropology*, **26**, 131–3.

Kaplan, M. (2002). Plight of the condor. *New Scientist*, 5 October, 34–6.

Kawai, M. (1965). Newly-acquired pre-cultural behavior of the natural troop of Japanese monkeys on Koshima Islet. *Primates*, **6**, 1–30.

Kawai, M., Watanabe, K. & Mori, A. (1992). Precultural behaviors observed in free-ranging Japanese monkeys on Koshima Islet over the past 25 years. *Primate Report*, **32**, 143–53.

Kawamoto, Y., Nozawa, K., Matsubayashi, K. & Gotoh, S. (1988). A population-genetic study of crab-eating macaques (*Macaca fascicularis*) on the island of Anguar, Palau, Micronesia. *Folia Primatologica*, **51**, 169–81.

Kawamura, S. (1959). The process of sub-culture propagation among Japanese macaques. *Primates*, **2**, 43–60.

Kawanaka, K. (1990). Alpha males' interactions and social skills. In: *The Chimpanzees of the Mahale Mountains*, ed. T. Nishida, Tokyo: University of Tokyo Press, pp. 171–87.

Kerbis Peterhans, J. C., Wrangham, R. W., Carter, M. L. & Hauser, M. D. (1993). A contribution to tropical rain forest taphonomy: retrieval and documentation of chimpanzee remains from Kibale Forest, Uganda. *Journal of Human Evolution*, **25**, 485–514.

Köhler, W. (1927). *The Mentality of Apes*, 2nd edn. London: Kegan Paul, Trench, Truber.

Kondo, M., Kawamoto, Y., Nozawa, K., Matsubayashi, K., Watanabe, T., Griffiths, O. & Stanley, M. A. (1993). Population genetics of crab-eating macaques (*Macaca fascicularis*) on the island of Mauritius. *American Journal of Primatology*, **29**, 167–82.

Kortlandt, A. (1962). Chimpanzees in the wild. *Scientific American*, **206**, 128–38.

 (1967). Experimentation with chimpanzees in the wild. In: *Neue Ergebnisse der Primatologie*, eds. D. Starck, R. Schneider & H-J. Kuhn, Stuttgart: Gustav Fischer, pp. 208–24.

 (1986). The use of stone tools by wild-living chimpanzees and earliest hominids. *Journal of Human Evolution*, **15**, 77–132.

Kortlandt, A. & Holzhaus, E. (1987). New data on the use of stone tools by chimpanzees in Guinea and Liberia. *Primates*, **28**, 473–96.

Krings, M., Stone, A., Schmitz, R. W., Krainitzki, H., Stoneking, M. & Pääbo, S. (1997). Neandertal DNA sequences and the origin of modern humans. *Cell*, **90**, 19–30.

Kroeber, A. L. (1928). Sub-human culture beginnings. *Quarterly Review of Biology*, **3**, 325–42.

Kroeber, A. L. & Kluckhohn, C. (1963). *Culture: A Critical Review of Concepts and Definitions.* New York: Random House.

Kummer, H. (1971). *Primate Societies. Group Techniques of Ecological Adaptation.* Chicago: Aldine-Atherton.

Kuper, A. (1999). *Culture. The Anthropologist's Account.* Cambridge, MA: Harvard University Press.

Kuroda, S., Nishihara, T., Suzuki, S. & Oko, R. A. (1996). Sympatric chimpanzees and gorillas in the Ndoki Forest, Congo. In: *Great Ape Societies,* eds. W. C. McGrew, L. F. Marchant & T. Nishida, Cambridge: Cambridge University Press, pp. 71–81.

Kusmierski, R., Borgia, G., Uy, A. & Crozier, R. H. (1997). Labile evolution of display traits in bowerbirds indicates reduced effects of phylogenetic constraint. *Proceedings of the Royal Society of London, B,* **264**, 307–13.

Kuznar, L. A. (1997). *Reclaiming a Scientific Anthropology.* Walnut Creek, CA: Altamira Press.

Ladygina-Kohts, N. N. (2002). *Infant Chimpanzee and Human Child.* New York: Oxford University Press.

Laland, K. N, & Hoppitt, W. (2003). Do animals have culture? *Evolutionary Anthropology,* **12**, 150–9.

Laland, K. N., Odling-Smee, J. & Feldman, M. W. (2000). Niche construction, biological evolution, and cultural change. *Behavioral and Brain Sciences,* **23**, 131–75.

Lancaster, J. B. (1975). *Primate Behavior and the Emergence of Human Culture.* New York: Holt, Rinehart & Winston.

Lanjouw, A. (2002). Behavioural adaptations to water scarcity in Tongo chimpanzees. In: *Behavioural Diversity in Chimpanzees and Bonobos,* eds. C. Boesch, G. Hohmann & L. F. Marchant, Cambridge: Cambridge University Press, pp. 52–60.

Lethmate, J. (1982). Tool-using skills of orangutans. *Journal of Human Evolution,* **11**, 49–64.

Linden, E. (2002) *The Octopus and the Orangutan.* New York: Dutton.

Logan, M. L. & Qirko, H. N. (1996). An evolutionary perspective on maladaptive traits and cultural conformity. *American Journal of Human Biology,* **8**, 615–29.

Lorenz, K. (1977). *Behind the Mirror. A Search for a Natural History of Human Knowledge.* London: Methuen.

Lumsden, C. J. & Wilson, E. O. (1981). *Genes, Mind, and Culture.* Cambridge, MA: Harvard University Press.

Lyon, B. E. (2003). Egg recognition and counting reduce costs of avian conspecific brood parasitism. *Nature,* **422**, 495–9.

Madden, J. R. (2003). Male spotted bowerbirds preferentially choose, arrange and proffer objects that are good predictors of mating success. *Behavioral Ecology and Sociobiology,* **148**, 421–52.

Maestripieri, D. (1996). Maternal encouragement of infant locomotion in pigtail macaques, *Macaca nemestrina. Animal Behaviour,* **51**, 603–10.

Manson, J. H. & Wrangham, R. W. (1991). Intergroup aggression in chimpanzees and humans. *Current Anthropology*, **32**, 369–80.

Maples, W. R., Maples, M. K., Greenhood, W. F. & Walek, M. L. (1976a). Adaptations of crop-raiding baboons in Kenya. *American Journal of Physical Anthropology*, **45**, 309–16.

Maples, W. R., Brown, A. B. & Hutchens, P. M. (1976b). Introduced monkey populations at Silver Springs, Florida. *Florida Anthropologist*, **29**, 133–6.

Marchant, L. F. (2002). *Chimpanzee Carnivory: Why Eat Meat?* Oxford, OH: Miami University Audiovisual Services. (VHS video.)

(2003). *Studying Wild Chimpanzees in Senegal*. Oxford, OH: Miami University Audiovisual Services. (VHS video.)

Marchant, L. F. & McGrew, W. C. (1999). Innovative behaviour at Mahale: new data on nasal probe and nipple press. *Pan Africa News*, **6**, 16–18.

(2004). Percussive technology: chimpanzee baobab smashing and the evolutionary modelling of hominid knapping. In: *Knapping Stones: A Uniquely Human Behaviour?* ed. V. Rous & B. Bril, forthcoming. Cambridge: McDonald Institute Monograph Series.

Marchesi, P., Marchesi, N., Fruth, B. & Boesch, C. (1995). Census and distribution of chimpanzees in Côte d'Ivoire. *Primates*, **36**, 591–607.

Marlar, R. A., Leonard, B. L., Billman, B. R., Lambert, P. M. & Marler, J. E. (2000). Biochemical evidence of cannibalism at a prehistoric Puebloan site in southwestern Colorado. *Nature*, **407**, 74–8.

Marler, P. (1969). Vocalizations of wild chimpanzees. An introduction. In: *Proceedings of the Second International Congress of Primatology*, Vol. I: *Behavior*, ed. C. R. Carpenter, Basel: S. Karger, pp. 94–100.

(1996). Social cognition. Are primates smarter than birds? In: *Current Anthropology*, Vol. 13, eds. V. Nolan & E. D. Ketterson, New York: Plenum, pp. 1–32.

Marler, P. & Hobbett, L. (1975). Individuality in a large-range vocalization of wild chimpanzees. *Zeitschrift für Tierpsychologie*, **32**, 97–109.

Marler, P. & Tamura, M. (1964). Culturally transmitted patterns of vocal behavior in sparrows. *Science*, **146**, 1483–6.

Marshall, A. J., Wrangham, R. W. & Arcadi, A. C. (1999). Does learning affect the structure of vocalizations in chimpanzees? *Animal Behaviour*, **58**, 825–30.

Matsuzawa, T. (1991). Nesting cups and metatools in chimpanzees. *Behavioral and Brain Sciences*, **14**, 570–1.

(1994). Field experiments on the use of stone tools by chimpanzees in the wild. In: *Chimpanzee Cultures*, eds. R. W. Wrangham, W. C. McGrew, F. B. M. de Waal & P. G. Heltne, Cambridge, MA: Harvard University Press, pp. 351–70.

(2003). Koshima monkeys and Bossou chimpanzees: long-term research on culture in non-human primates. In: *Animal Social Complexity: Intelligence, Culture, and Individualized*

Societies, eds. F. B. M. de Waal & P. L. Tyack, Cambridge, MA: Harvard University Press, pp. 374–87.

Matsuzawa, T., Takemoto, H., Hayakawa, S. & Shimada, M. (1999). Diecke Forest in Guinea. *Pan Africa News*, **6**, 10–11.

Matsuzawa, T. & Yamakoshi, G. (1996). Comparison of chimpanzee material culture between Bossou and Nimba, West Africa. In: *Reaching into Thought: The Mind of the Great Apes*, eds. A. E. Russon, K. A. Bard & S. T. Parker, Cambridge: Cambridge University Press, pp. 211–32.

Maughan, J. E. & Stanford, C. B. (2001). Terrestrial nesting by chimpanzees in Bwindi Impenetrable National Park, Uganda. *American Journal of Physical Anthropology*, Supplement 32, 104. (Abstract.)

McGrew, W. C. (1974). Tool use by wild chimpanzees in feeding upon driver ants. *Journal of Human Evolution*, **3**, 501–8.

(1977). Socialization and object manipulation of wild chimpanzees. In: *Primate Bio-Social Development*, eds. S. Chevalier-Skolnikoff & F. E. Porter, New York: Garland, pp. 261–88.

(1979). Evolutionary implications of sex differences in chimpanzee predation and tool use. In: *The Great Apes*, eds. D. A. Hamburg & E. R. McCown, Menlo Park, CA: Benjamin/Cummins, pp. 440–63.

(1981). The female chimpanzee as a human evolutionary prototype. In: *Woman the Gatherer*, ed. F. Dahlberg, New Haven, CT: Yale University Press, pp. 35–73.

(1987). Tools to get food: the subsistants of Tasmanian aborigines and Tanzanian chimpanzees compared. *Journal of Anthropological Research*, **43**, 247–58.

(1989). Why is ape tool use so confusing? In: *Comparative Socioecology: The Behavioural Ecology of Humans and Other Mammals*, eds. V. Standen & R. A. Foley, Cambridge: Cambridge University Press, pp. 457–72.

(1992). *Chimpanzee Material Culture. Implications for Human Evolution.* Cambridge: Cambridge University Press.

(1998). Culture in nonhuman primates? *Annual Review of Anthropology*, **27**, 301–28.

(2001a). The nature of culture: prospects and pitfalls of cultural primatology. In: *Tree of Origin. What Primate Behavior can Tell Us about Human Social Evolution*, ed. F. B. M. de Waal, Cambridge, MA: Harvard University Press, pp. 231–54.

(2001b). The other faunivory: primate insectivory and early human diet. In: *Meat-Eating and Human Evolution*, eds. C. B. Stanford & H. T. Bunn, Oxford: Oxford University Press, pp. 160–78.

(2003a). Evolution of the cultured mind: lessons from wild chimpanzees. In: *The Social Brain: Evolution and Pathology*, eds. M. Brüne, H. Ribbert & W. Schiefenhövel, New York: J. Wiley & Sons, pp. 81–92.

(2003b). Ten dispatches from the chimpanzee culture wars. In: *Animal Social Complexity. Intelligence, Culture, and Individualized Societies*, eds. F. B. M. de Waal & P. L. Tyack, Cambridge, MA: Harvard University Press, pp. 419–39.

McGrew, W. C., Baldwin, P. J., Marchant, L. F., Pruetz, J. D., Scott, S. E. & Tutin, C. E. G. (2003). Ethoarchaeology and elementary technology of unhabituated wild chimpanzees at Assirik, Senegal, West Africa. *Palaeoanthropology*, 1, 1–20.

McGrew, W. C., Baldwin, P. J. & Tutin, C. E. G. (1981). Chimpanzees in a hot, dry and open habitat: Mt. Assirik, Senegal, West Africa. *Journal of Human Evolution*, 10, 227–44.

McGrew, W. C., Ensminger, A. L., Marchant, L. F., Pruetz, J. D. & Vigilant, L. (2004). Genotyping aids field study of unhabituated wild chimpanzees. *American Journal of Primatology*, *forthcoming*.

McGrew, W. C., Ham, R. M., White, L. T. J., Tutin, C. E. G. & Fernandez, M. (1997). Why don't chimpanzees in Gabon crack nuts? *International Journal of Primatology*, 18, 353–74.

McGrew, W. C. & Marchant, L. F. (1998). Chimpanzee wears a knotted skin 'necklace'. *Pan Africa News*, 5, 8–9.

McGrew, W. C., Marchant, L. F., Scott, S. E. & Tutin, C. E. G. (2001). Intergroup differences in a social custom of wild chimpanzees: the grooming hand-clasp of the Mahale Mountains. *Current Anthropology*, 42, 148–53.

McGrew, W. C., & Marchant, L. F., Wrangham, R. W. & Klein, H. (1999). Manual laterality in anvil use: wild chimpanzees cracking *Strychnos* fruits. *Laterality*, 4, 79–87.

McGrew, W. C. & Tutin, C. E. G. (1978). Evidence for a social custom in wild chimpanzees. *Man*, 13, 234–51.

McGrew, W. C., Tutin, C. E. G. & Baldwin, P. J. (1979). Chimpanzees, tools and termites: cross-cultural comparisons of Senegal, Tanzania, and Rio Muni. *Man*, 14, 185–214.

McGrew, W. C., Tutin, C. E. G. & Midgett, P. S. (1975). Tool use in a group of captive chimpanzees. I. Escape. *Zeitschrift für Tierpsychologie*, 37, 145–62.

Menzel, E. W. (1972). Spontaneous invention of ladders in a group of young chimpanzees. *Folia Primatologica*, 17, 87–106.

(1973a). Further observations on the use of ladders in a group of young chimpanzees. *Folia Primatologica*, 19, 450–7.

(ed.) (1973b). *Precultural Primate Behavior*. Basel: S. Karger.

Mercader, J., Panger, M. A. & Boesch, C. (2002). Excavation of a chimpanzee stone tool site in the African rainforest. *Science*, 296, 1452–5.

Merfield, F. G. & Miller, H. (1956). *Gorillas Were My Neighbors*. London: Companion Book Club.

Mettke-Hofmann, C. & Gwinner, E. (2003). Long-term memory for a life on the move. *Proceedings of the National Academy of Sciences (USA)*, 100, 5863–6.

Mitani, J. C. & Gros-Louis, J. (1995). Species and sex differences in the screams of chimpanzees and bonobos. *International Journal of Primatology*, **16**, 339–411.

(1998). Chorusing and call convergence in chimpanzees: tests of three hypotheses. *Behaviour*, **135**, 1041–64.

Mitani, J. C., Gros-Louis, J. & Macedonia, T. M. (1996). Selection for acoustic individuality within the vocal repertoire of wild chimpanzees. *International Journal of Primatology*, **17**, 569–83.

Mitani, J. C. Hasegawa, T., Gros-Louis, J., Marler, P. & Byrne, R. W. (1992). Dialects in wild chimpanzees? *American Journal of Primatology*, **27**, 233–43.

Mitani, J. C., Hurley, K. L. & Murdoch, M. S. (1999). Geographic variation in the calls of wild chimpanzees: a reassessment. *American Journal of Primatology*, **47**, 133–51.

Mitani, J. C. & Stuht, J. (1998). The evolution of nonhuman primate loud calls: acoustic adaptation for long-distance transmission. *Primates*, **39**, 171–82.

Mitani, J. C. & Watts, D. P. (1999). Demographic influences on the hunting behavior of chimpanzees. *American Journal of Physical Anthropology*, **109**, 439–54.

Moore, J. (1985). Chimpanzee survey in Mali, West Africa. *Primate Conservation*, **6**, 59–63.

(1996). Savanna chimpanzees, referential models and the last common ancestor. In: *Great Ape Societies*, eds. W. C. McGrew, L. F. Marchant & T. Nishida, Cambridge: Cambridge University Press, pp. 275–92.

Morgan, D. & Sanz, C. (2003). Naïve encounters with chimpanzees in the Goualougo triangle, Republic of Congo. *International Journal of Primatology*, **24**, 369–81.

Morgan, L. H. (1868). *The American Beaver and His Works*. Philadelphia: J. P. Lippincott.

Morin, P. A., Moore, J. J., Chakraborty, R., Jin, L., Goodall, J. & Woodruff, D. S. (1994). Kin selection, social structure, gene flow, and the evolution of chimpanzees. *Science*, **265**, 1193–201.

Morris, D. (1981). *The Soccer Tribe*. London: Weidenfeld & Nicolson.

Morris, K. & Goodall, J. (1977). Competition for meat between chimpanzees and baboons of the Gombe National Park. *Folia Primatologica*, **28**, 109–21.

Muller, M. N. (2002). Agonistic relations among Kanyawara chimpanzees. In: *Behavioural Diversity in Chimpanzees and Bonobos*, eds. C. Boesch, G. Hohmann & L. F. Marchant, Cambridge: Cambridge University Press, pp. 112–24.

Mundinger, P. C. (1980). Animal cultures and a general theory of cultural evolution. *Ethology and Sociobiology*, **1**, 183–223.

Murdock, G. P., Ford, C. S., Hudson, A. E., Kennedy, R., Simmons, L. W. & Whiting, J. W. M. (1965). *Outline of Cultural Mate*rials. 4th edn. New Haven, CT: Human Relations Area Files, Inc.

Murdock, G. P. & Morrow, D. O. (1970). Subsistence economy and supportive practices: cross-cultural codes 1. *Ethnology*, **9**, 302–30.

Murdock, G. P. & Provost, C. (1973). Measurement of cultural complexity. *Ethnology*, **12**, 379–92.

Murray, S. S., Schoeninger, M. S., Bunn, H. T., Pickering, T. R. & Marlett, J. A. (2001). Nutritional composition of some wild plant foods and honey used by Hadza foragers of Tanzania. *Journal of Food Composition and Analysis*, **14**, 3–13.

Nakamichi, M., Kato, E., Kojima, Y. & Itoigawa, N. (1998). Carrying and washing of grass roots by free-ranging Japanese macaques at Katsuyama. *Folia Primatologica*, **69**, 35–40.

Nakamura, M. (2002). Grooming hand-clasp in Mahale M group chimpanzees: implications for culture in social behaviours. In: *Behavioural Diversity in Chimpanzees and Bonobos*, eds. C. Boesch, G. Hohmann & L. F. Marchant, Cambridge: Cambridge University Press, pp. 71–83.

(2003). 'Gatherings' of social grooming among wild chimpanzees: implications for evolution of sociality. *Journal of Human Evolution*, **44**, 59–71.

Nakamura, M., McGrew, W. C., Marchant, L. F. & Nishida, T. (2000). Social scratch: another custom in wild chimpanzees? *Primates*, **41**, 237–48.

Nakamura, M. & Nishida, T. (2002). Molding in wild chimpanzees: don't they really teach? [*Primate Research*], **18**, 373. (Abstract in Japanese.)

Nakamura, M. & Uehara, S. (2004). Proximate factors of different types of grooming-hand-clasp in Mahale chimpanzees: implications for chimpanzee social customs. *Current Anthropology*, **45**, 108–14.

Nishida, T. (1973). The ant-gathering behaviour by the use of tools among wild chimpanzees of the Mahale Mountains. *Journal of Human Evolution*, **2**, 357–70.

(1980a). Local differences in responses to water among wild chimpanzees. *Folia Primatologica*, **33**, 189–209.

(1980b). The leaf-clipping display: a newly-discovered expressive gesture in wild chimpanzees. *Journal of Human Evolution*, **9**, 117–28.

(1983). Alpha status and agonistic alliance in wild chimpanzees (*Pan troglodytes schweinfurthii*). *Primates*, **24**, 318–36.

(ed.) (1990). *The Chimpanzees of the Mahale Mountains: Sexual and Life History Strategies*. Tokyo: University of Tokyo Press.

(1993). Chimpanzees are always new to me. In: *The Great Ape Project. Equality beyond Humanity*, eds. P. Cavilieri & P. Singer, London: Fourth Estate, pp. 24–6.

(1994). Review of recent findings on Mahale chimpanzees. In: *Chimpanzee Cultures*, eds. R. W. Wrangham, W. C. McGrew, F. B. M. de Waal & P. Heltne, Cambridge, MA: Harvard University Press, pp. 373–96.

Nishida, T., Hasegawa, T., Hayaki, H., Takahata, Y. & Uehara, S. (1992). Meat-sharing as a coalition strategy by an alpha male chimpanzee? In: *Topics in Primatology*, Vol. 1: *Human Origins*, eds. T. Nishida, W. C. McGrew, P. Marler, M. Pickford & F. B. M. de Waal, Tokyo: University of Tokyo Press, pp. 159–74.

Nishida, T. & Hiraiwa, M. (1982). Natural history of a tool-using behaviour by wild chimpanzees in feeding on wood-boring ants. *Journal of Human Evolution*, **11**, 73–99.

Nishida, T., Hosaka, K., Nakamura, M. & Hamai, M. (1995). A within-group attack on a young adult male chimpanzee: ostracism of an ill-mannered member? *Primates*, **36**, 207–11.

Nishida, T., Kano, T., Goodall, J., McGrew, W. C. & Nakamura, M. (1999). Ethogram and ethnography of Mahale chimpanzees. *Anthropological Science*, **107**, 141–88.

Nishida, T. & Kawanaka, K. (1985). Within-group cannibalism by adult male chimpanzees. *Primates*, **26**, 274–84.

Nishida, T. Mitani, J. C. & Watts, D. P. (2004). Variable grooming behaviours in wild chimpanzees. *Folia Primatologica*, **75**, 31–6.

Nishida, T. & Nakamura, M. (1993). Chimpanzee's tool use to clear a blocked nasal passage. *Folia Primatologica*, **61**, 218–20.

Nishida, T., Uehara, S. & Ramadhani, N. (1979). Predatory behavior among wild chimpanzees of the Mahale Mountains. *Primates*, **20**, 1–20.

Nishida, T. & Wallauer, W. (2003). Leaf-pile pulling: an unusual play pattern in wild chimpanzees. *American Journal of Primatology*, **60**, 167–73.

Nishida, T., Wrangham, R. W., Goodall, J. & Uehara, S. (1983). Local differences in plant-feeding habits of chimpanzees between the Mahale Mountains and Gombe National Park. *Journal of Human Evolution*, **12**, 467–80.

Nissen, H. W. (1931). A field study of the chimpanzee: observations of chimpanzee behavior and environment in western French Guinea. *Comparative Psychology Monographs*, **8**, 1–122.

Noad, M. J., Cato, D. H., Bryden, M. M., Jenner, M. N. & Jenner, K. C. S. (2000). Cultural revolution in whale songs. *Nature*, **408**, 537.

Norton-Griffiths, M. (1967). Some ecological aspects of the feeding behaviour of the oystercatcher (*Haematopus ostralegus*) on the edible mussel *Mytilus edulis*. *Ibis*, **109**, 412–24.

O'Malley, R. C. & McGrew, W. C. (2000). Oral tool use by captive orangutans (*Pongo pygmaeus*). *Folia Primatologica*, **71**, 334–41.

Oswalt, W. H. (1976). *An Anthropology Analysis of Food-Getting Technology*. New York: John Wiley.

Oxford, P. (2003). Cracking monkeys. *BBC Wildlife*, **21**(2), 26–9.

Panger, M. A., Brooks, A. S., Richmond, B. G. & Wood, B. (2002a). Older than the Oldowan? Rethinking the emergence of hominin tool use. *Evolutionary Anthropology*, **11**, 235–45.

Panger, M. A., Perry, S., Rose, L., Gros-Louis, J., Vogel, E., MacKinnon, K. C. & Baker, M. (2002b). Cross-site differences in foraging behavior of white-faced capuchins. *American Journal of Physical Anthropology*, **119**, 52–66.

Parnell, R. J. & Buchanan-Smith, H. M. (2001). An unusual social display by gorillas. *Nature*, **412**, 294.

Payne, R. & McVay, S. (1971). Songs of humpback whales. *Science*, **173**, 587–97.

Perry, S., Baker, M., Fedigan, L., Gros-Louis, J., Jack, K., MacKinnon, K. C., Manson, J. H., Panger, M., Pyle, K. & Rose, L. (2003a). Social conventions in wild white-faced capuchin monkeys: evidence for behavioral traditions in a neotropical primate. *Current Anthropology*, **44**, 241–68.

Perry, S. & Manson, J. H. (2003). Traditions in monkeys. *Evolutionary Anthropology*, **12**, 71–81.

Perry, S., Panger, M., Rose, L. M., Baker, M., Gros-Louis, J., Jack, K., MacKinnon, K. C., Manson, J., Fedigan, L. & Pyle, K. (2003b). Traditions in wild white-faced capuchin monkeys. In: *The Biology of Traditions: Models and Evidence*, eds. D. M. Fragaszy, & S. Perry, Cambridge: Cambridge University Press, pp. 391–425.

Perry, S. & Rose, L. M. (1994). Begging and transfer of meat by white-faced capuchin monkeys, *Cebus capucinus*. *Primates*, **35**, 409–15.

Phillips, K. A. (1998). Tool use in wild capuchin monkeys (*Cebus albifrons trinitatis*). *American Journal of Primatology*, **46**, 259–61.

Pickering, T. & Wallis, J. (1997). Bone modifications resulting from captive chimpanzee mastication: implications for the interpretation of Pliocene archaeological faunas. *Journal of Archaeological Science*, **24**, 1115–27.

Plummer, T. & Stanford, C. B. (2000). Analysis of a prey bone assemblage made by wild chimpanzees in Gombe National Park, Tanzania. *Journal of Human Evolution*, **39**, 245–65.

Power, M. (1991). *The Egalitarians. Human and Chimpanzee: An Anthropological View of Social Organization*. New York: Cambridge University Press.

Premack, D. & Premack, A. J. (1994). Why animals have neither culture nor history. In: *Companion Encyclopedia of Anthropology*, ed. T. Ingold, London: Routledge.

Price, D. H. (1990). *Atlas of World Cultures. A Geographical Guide to Ethnographic Literature*. Newbury Park, CA: Sage Publications.

Pruetz, J. D. (2001). Use of caves by savanna chimpanzees (*Pan troglodytes verus*) in the Tomboronkoto region of southeastern Senegal. *Pan Africa News*, **8**, 26–8.

Pruetz, J. D., Marchant, L. F., Arno, J. & McGrew, W. C. (2002). Survey of savanna chimpanzees (*Pan troglodytes verus*) in southeastern Senegal. *American Journal of Primatology*, **58**, 35–43.

Pryor, K. W., Haag, R. & O'Reilly, T. (1969). The creative porpoise: training for novel behavior. *Journal of the Experimental Analysis of Behavior*, **12**, 653–71.

Pryor, K. W., Lindbergh, J., Lindbergh, S. & Milano, R. (1990). A dolphin–human fishing cooperative in Brazil. *Marine Mammal Science*, **6**, 77–82.

Ramsey, J. K. & McGrew, W. C. (2004). Object play in great apes. In: *The Nature of Play: Great Apes and Humans*, eds. A. D. Pellegrini & P. K. Smith, New York: Guildford Press, *forthcoming*.

Rawlins, R. G. & Kessler, M. J. (eds.) (1986). *The Cayo Santiago Macaques: History, Behavior and Biology*. Albany: State University of New York Press.

Reader, S. M. & Laland, K. (2001). Primate innovation: sex, age and social rank differences. *International Journal of Primatology*, **22**, 787–805.

Relethford, J. H. (2000). *The Human Species: An Introduction to Biological Anthropology*, 4th edn. Mountain View, CA: Mayfield.

Rendell, L. & Whitehead, H. (2001). Culture in whales and dolphins. *Behavioral and Brain Sciences*, **24**, 309–82.

(2003). Vocal clans in sperm whales (*Physeter macrocephalus*). *Proceedings of the Royal Society of London B*, **270**, 222–31.

Reynolds, V. (1992). Chimpanzees of the Budongo Forest, 1962–1992. *Journal of Zoology*, **228**, 695–9.

Reznikova, Z. (2001). Interspecific and intraspecific social learning in ants. *Advances in Ethology*, **36**, 108 (abstract).

Robbins, M. M. (2001). Variation in the social system of mountain gorillas: the male perspective. In: *Mountain Gorillas: Three Decades of Research at Karisoke*, eds. M. M. Robbins, P. Sicotte, K. J. Stewart, New York: Cambridge University Press, pp. 29–58.

Rose, L. M. (1994). Sex differences in diet and foraging behavior in white-faced capuchins (*Cebus capucinus*). *International Journal of Primatology*, **15**, 63–82.

(1997). Vertebrate predation and food-sharing in *Cebus* and *Pan*. *International Journal of Primatology*, **18**, 727–65.

Sakura, O. & Matsuzawa, T. (1991). Flexibility of wild chimpanzee nut-cracking behavior using stone hammers and anvils: an experimental analysis. *Ethology*, **87**, 237–48.

Savage, T. S. & Wyman, J. (1844). Observations on the external characters and habits of *Troglodytes niger*, Geoff. and on its organization. *Boston Journal of Natural History*, **4**, 362–86.

Savage-Rumbaugh, E. S. (1986). *Ape Language: From Conditioned Response To Symbol*. New York: Columbia University Press.

van Schaik, C. P. (1999). The socioecology of fission–fusion sociality in orangutans. *Primates*, **40**, 73–90.

(2000). Infanticide by male primates: the sexual selection hypothesis revisited. In: *Infanticide in Males and its Implications*, eds. C. P. van Schaik & C. H. Janson, Cambridge: Cambridge University Press, pp. 27–60.

(2003). Local traditions in orangutans and chimpanzees: social learning and social tolerance. In: *The Biology of Traditions: Models and Evidence*, eds. D. M. Fragaszy, & S. Perry, Cambridge: Cambridge University Press, pp. 297–328.

REFERENCES 217

van Schaik, C. P., Ancrenaz, M., Borgen, G., Galdikas, B., Knott, C. D., Singleton, I., Suzuki, A., Utami, S. S. & Merrill, M. (2003). Orangutan cultures and the evolution of material culture. *Science*, **299**, 102–5.

van Schaik, C. P., Fox, E. A. & Sitompul, A. F. (1996). Manufacture and use of tools in wild Sumatran orangutans: implications for human evolution. *Naturwissenschaften*, **83**, 186–8.

van Schaik, C. P. & Knott, C. D. (2001). Geographic variation in tool use on *Neesia* fruits in orangutans. *American Journal of Physical Anthropology*, **114**, 331–42.

Schick, K. D. & Toth, N. (1993). *Making Silent Stones Speak. Human Evolution and the Dawn of Technology.* New York: Simon & Schuster.

Schick, K. A., Toth, N., Garufi, G., Savage-Rumbaugh, E. S., Rumbaugh, D. & Sevcik, R. A. (1999). Continuing investigations into the stone tool-making and tool-using capabilities of bonobo (*Pan paniscus*). *Journal of Archaeological Science*, **26**, 821–32.

Schiefenhövel, W. (1989). Reproduction and sex-ratio manipulation through preferential female infanticide among the Eipo, in the highlands of West-New Guinea. In: *The Sociobiology of Sexual and Reproductive Strategies*, eds. A. Rosa, C. Vogel & E. Voland, London: Chapman & Hall, pp. 170–93.

Shweder, R. A. (2001). Rethinking the object of anthropology and ending up where Kroeber and Kluckhohn began. *American Anthropologist*, **103**, 437–40.

Semaw, S., Renne, P., Harris, J. W. K., Feibel, C. S., Bernor, R. L., Fesseha, N. & Mowbray, K. (1996). 2.5 million-year-old stone tools from Gona, Ethiopia. *Nature*, **385**, 333–8.

Sept, J. M. (1992). Was there no place like home? A new perspective on early hominid archaeological sites from the mapping of chimpanzee nests. *Current Anthropology*, **33**, 187–207.

(1998). Shadows on a changing landscape: comparing nesting habits of hominids and chimpanzees since their last common ancestor. *American Journal of Primatology*, **46**, 85–101.

Shimada, M. (2003). A note on the southern neighboring groups of M group in the Mahale Mountains National Park. *Pan Africa News*, **10**(1), 11–14.

Sidky, H. (2003). *A Critique of Postmodern Anthropology – In Defense of Disciplinary Origins and Traditions.* Lewiston: Edwin Mellen Press.

Simões-Lopes, P. C., Fabian, M. E. & Menegheti, J. O. (1998). Dolphin interactions with the mullet artisanal fishing on southern Brazil: a qualitative and quantitative approach. *Revista Brasiliera de Zoologica*, **15**, 709–26.

Slater, P. J. B. (1986). The cultural transmission of bird song. *Trends in Ecology and Evolution*, **1**, 94–7.

Smith, E. A., Mulder, M. B. & Hill, K. (2001). Controversies in the evolutionary social sciences: a guide for the perplexed. *Trends in Ecology and Evolution*, **16**, 128–35.

Smolker, R. A., Richards, A. F., Connor, R. C., Mann, J. & Berggren, P. (1997). Sponge-carrying by Indian Ocean bottlenose dolphins: possible tool-use by a delphinid. *Ethology*, **103**, 454–65.

Snowdon, C. T. (2001). From primate communication to human language. In: *Tree of Origin. What Primate Behavior Can Tell Us About Human Social Evolution*, ed. F. B. M. de Waal, pp. 193–227, Cambridge, MA: Harvard University Press.

Sommer, V. (2000). The holy wars about infanticide. Which side are you on? And why? In: *Infanticide by Males and its Implications*, eds. C. P. van Schaik & C. H. Janson, Cambridge: Cambridge University Press, pp. 9–26.

Sommer, V., Fowler, A., Adanu, J. (2003). The Nigerian chimpanzee (*Pan troglodytes vellerosus*) at Gashaka: two years of habituation efforts. *Folia Primatologica*, **74**, 222 (abstract).

Stanford, C. B. (1998a). *Chimpanzee and Red Colobus: The Ecology of Predator and Prey*. Cambridge, MA: Harvard University Press.

(1998b). The social behavior of chimpanzees and bonobos: empirical evidence and shifting assumptions. *Current Anthropology*, **39**, 399–420.

(2002). Arboreal bipedalism in Bwindi chimpanzees. *American Journal of Physical Anthropology*, **119**, 87–91.

Stokes, E. J. & Byrne, R. W. (2001). Cognitive capacities for behavioural flexibility in wild chimpanzees (*Pan troglodytes*): the effect of snare injury on complex mammal food processing. *Animal Cognition*, **4**, 11–28.

Strassman, J. E., Zhu, Y. & Queller, D. C. (2000). Altruism and social cheating in the social amoeba (*Dictyostelium discoideum*). *Nature*, **408**, 965–7.

Strier, K. B. (2003). Primate behavioral ecology: from ethnography to ethology and back. *American Anthropologist*, **105**, 16–27.

Struhsaker, T. T. & Hunkeler, P. (1971). Evidence of tool-using by chimpanzees in the Ivory Coast. *Folia Primatologica*, **15**, 212–19.

Sugiyama, Y. (1985). The brush-stick of chimpanzees found in south-west Cameroon and their cultural characteristics. *Primates*, **26**, 361–74.

(1995a). Drinking tools of wild chimpanzees at Bossou. *American Journal of Primatology*, **37**, 363–9.

(1995b). Tool-use for catching ants by chimpanzees at Bossou and Monts Nimba. *Primates*, **36**, 193–205.

(1997). Social tradition and the use of tool-composites by wild chimpanzees. *Evolutionary Anthropology*, **6**, 23–7.

Sugiyama, Y., Fushimi, T., Sakura, O. & Matsuzawa, T. (1993). Hand preference and tool use in wild chimpanzees. *Primates*, **34**, 151–9.

Sugiyama, Y. & Koman, J. (1979a). Social structure and dynamics of wild chimpanzees at Bossou, Guinea. *Primates*, **20**, 323–39.

(1979b). Tool-using and -making behavior in wild chimpanzees at Bossou, Guinea. *Primates*, **20**, 513–24.

Sugiyama, Y., Koman, J. & Sow, M. B. (1988). Ant-catching wands of wild chimpanzees at Bossou, Guinea. *Folia Primatologica*, **51**, 56–60.

Suzuki, A. (1965). An ecological study of wild Japanese monkeys in snowy areas-focused on their food habits-. *Primates*, **6**, 31–72.

(1969). An ecological study of chimpanzees in a savanna woodland. *Primates*, **10**, 103–48.

Suzuki, S., Hill, D. A., Marahashi, T. & Tsukahara, T. (1990). Frog and lizard-eating behaviour of wild Japanese macaques in Yakashima, Japan. *Primates*, **31**, 421–6.

Tappen, M. & Wrangham, R. W. (2000). Recognizing hominid-modified bones: the taphonomy of colobus bones partially digested by free-ranging chimpanzees in the Kibale Forest, Uganda. *American Journal of Physical Anthropology*, **113**, 217–34.

Tebbich, S., Taborsky, M., Fessl, B. & Blomqvist, D. (2001). Do woodpecker finches acquire tool-use by social learning? *Proceedings of the Royal Society of London B*, **268**, 2189–93.

Tebbich, S., Taborsky, M., Fessl, B. & Dvorak, M. (2002). The ecology of tool-use in the woodpecker finch (*Cactospiza pallida*). *Ecology Letters*, **5**, 656–64.

Temerlin, M. K. (1975). *Lucy: Growing Up Human*. Palo Alto, CA: Science and Behavior Books.

Terkel, J. (1996). Cultural transmission of feeding behavior in the black rat (*Rattus rattus*). In: *Social Learning in Animals: The Roots of Culture*, eds. C. M. Heyes, & B. G. Galef, San Diego, CA: Academic Press, pp. 17–47.

Thomas, D. H. (2000). *Skull Wars. Kennewick Man, Archaeology, and the Battle for Native American Identity*. New York: Basic Books.

Thompson, J. A. M. (2001). The status of bonobos in their southernmost geographic range. In: *All Apes Great and Small*. Vol. 1: *African Apes*, eds. B. M. F. Galdikas, N. E. Briggs, L. K. Sheeran, G. L. Shapiro & J. Goodall, New York: Kluwer Academic/Plenum, pp. 75–81.

(2002). Bonobos of the Lukuru Wildlife Research Project. In: *Behavioural Diversity in Chimpanzees and Bonobos*, eds. C. Boesch, G. Hohmann & L. F. Marchant, Cambridge: Cambridge University Press, pp. 61–70.

Tinbergen, N. (1951). *The Study of Instinct*. Oxford: Oxford University Press.

(1953). *Social Behaviour in Animals*. London: Chapman & Hall.

(1963). On aims and methods in ethology. *Zeitschrift für Tierpsychologie*, **20**, 410–33.

Tomasello, M. (1999). The human adaptation for culture. *Annual Review of Anthropology*, **28**, 509–29.

Tomasello, M. & Call, J. (1997). *Primate Cognition*. New York: Oxford University Press.

Tomasello, M., Call, J., Nagell, K., Olguin, R. & Carpenter, M. (1994). The learning and use of gestural signals by young chimpanzees. A transgenerational study. *Primates*, **35**, 137–54.

Tomasello, M., Kruger, A. C. & Ratner, H. H. (1993). Cultural learning. *Behavioral and Brain Sciences*, **16**, 495–552.

Toth, N., Schick, K. A., Savage-Rumbaugh, E. S., Sevcik, R. A., & Rumbaugh, D. M. (1993). *Pan* the tool-maker: investigations into the stone tool-making and tool-using capabilities of a bonobo (*Pan paniscus*). *Journal of Archaeological Science*, **20**, 81–91.

Tramo, M. J. (2001). Music of the hemispheres. *Science*, **291**, 54–6.

Trivers, R. (1985). *Social Evolution*. Menlo Park, CA: Benjamin/Cummings.

Tutin, C. E. G. (1979). Mating patterns and reproductive strategies in a community of wild chimpanzees (*Pan troglodytes schweinfurthii*). *Behavioral Ecology and Sociobiology*, **6**, 29–38.

Tutin, C. E. G., Ancrenaz, M. Paredes, J., Vachev-Vallas, M., Vidal, C., Goossens, B., Bruford, M. W. & Jamart, A. (2001). Conservation biology framework for the release of wild-born orphaned chimpanzees into the Conkouati Reserve, Congo. *Conservation Biology*, **15**, 1247–57.

Tutin, C. E. G. & Fernandez, M. (1985). Foods consumed by sympatric populations of *Gorilla g. gorilla* and *Pan t. trogolodytes* in Gabon: some preliminary data. *International Journal of Primatology*, **6**, 27–43.

Tutin, C. E. G., Fernandez, M., Rogers, M. E., Williamson, E. A. & McGrew, W. C. (1991). Foraging profiles of sympatric lowland gorillas and chimpanzees in the Lopé Reserve, Gabon. *Philosophical Transactions of the Royal Society, B*, **334**, 19–26.

Tutin, C. E. G. & Oslisly, R. (1995). *Homo*, *Pan*, and *Gorilla*: co-existence over 60,000 years at Lopé in central Gabon. *Journal of Human Evolution*, **28**, 597–602.

Tylor, E. B. (1871). *Primitive Culture*. London: Murray.

Vigilant, L. (2002). Technical challenges in the microsatellite genotyping of a wild chimpanzee group. *Evolutionary Anthropology*, Suppl. 1, 162–5.

Visalberghi, E. & Fragaszy, D. (2002). 'Do monkeys ape?' – ten years after. In: *Imitation in Animals and Artifacts*, eds. K. Dautenhohn & C. L. Nehaniv, Cambridge, MA: MIT Press, pp. 471–99.

Visalberghi, E. & McGrew, W. C. (1997). *Cebus* meets *Pan*. *International Journal of Primatology*, **18**, 677–81.

de Waal, F. B. M. (1982). *Chimpanzee Politics*. London: Jonathan Cape.

(1991). The chimpanzee's sense of social regularity and its relation to the human sense of justice. *American Behavioral Scientist*, **34**, 335–49.

(2001). *The Ape and the Sushi Master. Cultural Reflections by a Primatologist*. New York: Basic Books.

de Waal, F. B. M. & Lanting, F. (1997). *Bonobo: The Forgotten Ape*. Berkeley, CA: University of California Press.

de Waal, F. B. M. & Seres, M. (1997). Propagation of handclasp grooming among captive chimpanzees. *American Journal of Primatology*, **43**, 339–46.

Wallis, J. (1997). Chimpanzee consortships: new information on conception rate, seasonality, and individual preference. *American Journal of Primatology*, **42**, 152–3. (Abstract.)

Wallis, J. & Lee, D. R. (1999). Primate conservation: the prevention of disease transmission. *International Journal of Primatology*, **20**, 803–26.

Walsh, P. D., Abernethy, K. A., Bermejo, M., Beyers, R., de Wachter, P. Akou, M. E., *et al*. (2003). Catastrophic ape decline in western equatorial Africa. *Nature*, **422**, 611–14.

Warner, R. R. (1988). Traditionality of mating-site preferences in a coral reef fish. *Nature*, **335**, 719–21.

Washburn, S. L. & Benedict, B. (1979). Non-human primate culture. *Man*, **14**, 163–4.

Watanabe, K. (1989). Fish: a new addition to the diet of Japanese macaques on Koshima Island. *Folia Primatologica*, **52**, 124–31.

(1994). Precultural behavior of Japanese macaques: longitudinal studies of the Koshima troops. In: *The Ethological Roots of Culture*, eds. R. A. Gardner, B. T. Gardner, B. Chiarelli, & F. X. Plooij, Dordrecht: Kluwer, pp. 81–94.

(2001). A review of 50 years of research on the Japanese monkeys of Koshima: status and dominance. In: *Primate Origins of Human Cognition and Behavior*, ed. T. Matsuzawa, Tokyo: Springer, pp. 405–17.

Watts, D. P. (1989). Ant eating behavior of mountain gorillas. *Primates*, **30**, 121–5.

Watts, D. P. & Mitani, J. C. (2001). Boundary patrols and intergroup aggression in wild chimpanzees. *Behaviour*, **138**, 299–327.

(2002a). Hunting behavior of chimpanzees at Ngogo, Kibale National Park, Uganda. *International Journal of Primatology*, **23**, 1–28.

(2002b). New cases of inter-community infanticide by male chimpanzees at Ngogo, Kibale National Park, Uganda. *Primates*, **43**, 263–70.

Weinrich, M. T., Schilling, M. R. & Belt, C. R. (1992). Evidence for acquisition of a novel feeding behaviour: lobtail feeding in humpback whales, *Megaptera novaeangliae*. *Animal Behaviour*, **44**, 1059–72.

West, M. J., King A. P. & White, D. J. (2003). Discovering culture in birds: the role of learning and development. In: *Animal Social Complexity. Intelligence, Culture, and Individualized Societies*, eds. F. B. M. deWaal & P. L. Tyack, Cambridge, MA: Harvard University Press, pp. 470–91.

Westergaard, G. C. (1994). The subsistence technology of capuchins. *International Journal of Primatology*, **15**, 899–906.

(1995). The stone-tool technology of capuchin monkeys: possible implications for the evolution of symbolic communication in hominids. *World Archaeology*, **27**, 1–9.

(1998). What capuchin monkeys can tell us about the origins of hominid material culture. *Journal of Material Culture*, **3**, 5–19.

Westergaard, C. G. & Suomi, S. J. (1993). Use of a tool-set by capuchin monkeys (*Cebus apella*). *Primates*, **34**, 459–62.

(1995). The manufacture and use of bamboo tools by monkeys: possible implications for the development of material culture among East Asian hominids. *Journal of Archaeological Science*, **22**, 677–81.

Wheatley, B. D. (1999). *The Sacred Monkeys of Bali*. Prospect Heights, IL: Waveland Press.

White, T. D. (2001). Once were cannibals. *Scientific American*, August, 48–55.

White, T. D. & Suwa, G. (1987). Hominid footprints at Laetoli: facts and interpretations. *American Journal of Physical Anthropology*, **72**, 485–514.

Whitehead, H. (1998) Cultural selection and genetic diversity in matrilineal whales. *Science*, **282**, 1708–11.

Whiten, A. (1999). Parental encouragement in *Gorilla* in comparative perspective: implications for social cognition and the evolution of teaching. In: *The Mentalities of Gorillas and Orangutans: Comparative Perspectives*, eds. S. T. Parker, R. W. Mitchell & H. L. Miles, Cambridge: Cambridge University Press, pp. 342–66.

(2000). Primate culture and social learning. *Cognitive Science*, **24**, 477–508.

Whiten, A., Goodall, J., McGrew, W. C., Nishida, T., Reynolds, V., Sugiyama, Y. *et al.* (1999). Cultures in chimpanzees. *Nature*, **399**, 682–5.

(2001). Charting cultural variation in chimpanzees. *Behaviour*, **138**, 1481–516.

Whiten, A. & Ham, R. (1992). On the nature and evolution of imitation in the animal kingdom: reappraisal of a century of research. *Advances in the Study of Behavior*, **21**, 239–83.

Whiten, A., Horner, V., Litchfield, C. A. & Marshall-Pescini, S. (2003a). How do apes ape? *Animal Learning and Behaviour*, **32** (1), 36–52.

Whiten, A., Horner, V. & Marshall-Pescini, S. (2003b). Cultural panthropology. *Evolutionary Anthropology* **12**, 92–105.

Whitesides, G. H. (1985). Nut cracking by wild chimpanzees in Sierra Leone, West Africa. *Primates*, **26**, 91–4.

Whitfield, J. (2002). Nosy neighbours. *Nature*, **419**, 242–3.

Williamson, E. A. & Feistner, A. C. (2003). Habituating primates: processes, techniques, variables and ethics. In: *Field and Laboratory Methods in Primatology: A Practical Guide*, ed. J. M. Setchell, pp. 25–39. Cambridge: Cambridge University Press.

Winick, C. (1960). *Dictionary of Anthropology*. London: Peter Owen.

Wrangham, R. W. (1974). Artificial feeding of chimpanzees and baboons in their natural habitat. *Animal Behaviour*, **22**, 83–93.

(1977). Feeding behaviour of chimpanzees in Gombe National Park, Tanzania. In: *Primate Ecology: Studies of Feeding and Ranging Behaviour in Lemurs, Monkeys a*nd *Apes*, ed. T. H. Clutton-Brock, London: Academic Press, pp. 503–38.

(1999). Evolution of coalitionary killing. *Yearbook of Physical Anthropology*, **42**, 21–30.

Wrangham, R. W., Chapman, C. A., Clark-Arcadi, A. & Isabirye- Basuta, G. (1996). Social ecology of Kanyawara chimpanzees: implications for understanding the costs of great ape groups. In: *Great Ape Societies*, eds. W. C. McGrew, L. F. Marchant & T. Nishida, Cambridge: Cambridge University Press, pp. 45–57.

Wrangham, R. W., Hagel, G., Leighton, M., Marshall, A. J., Waldau, P. & Nishida, T. (2003). The Great Ape World Heritage Species Project. In: *Conservation in the 21st Century: Gorillas as a Case Study*, New York: Kluwer Academic/Plenum.

Wrangham, R. W. & Nishida, T. (1983). *Aspilia* spp. leaves: a puzzle in the feeding behavior of wild chimpanzees. *Primates*, **24**, 276–82.

Wrangham, R. W., de Waal, F. B. M. & McGrew, W. C. (1994). The challenge of behavioral diversity. In: *Chimpanzee Cultures*, eds. R. W. Wrangham, W. C. McGrew, F. B. M. de Waal & P. G. Heltne, Cambridge, MA: Harvard University Press, pp. 1–18.

Wynn, T. G. & McGrew, W. C. (1989). An ape's view of the Oldowan. *Man*, **24**, 383–98.

Yamakoshi, G. (2001). Ecology of tool use in wild chimpanzees: toward reconstruction of early hominid evolution. In: *Primate Origins of Human Cognition and Behavior*, ed. T. Matsuzawa, Tokyo: Springer, pp. 537–56.

Yamakoshi, G. & Sugiyama, Y. (1995). Pestle-pounding behavior of wild chimpanzees at Bossou, Guinea: a newly observed tool-using behavior. *Primates*, **36**, 489–500.

Yamagiwa, J. (1998). An ossified chimpanzee found in a tree nest. *Pan Africa News*, **5**, 17–18.

Yamagiwa, J., Muruhashi, T., Yumoto, T. & Mwanza, N. (1996). Dietary and ranging overlap in sympatric gorillas and chimpanzees in the Kahuzi-Biega National Park, Zaire. In: *Great Ape Societies*, eds. W. C. McGrew, L. F. Marchant & T. Nishida, Cambridge: Cambridge University Press, pp. 82–98.

Author index

Subject index

Where unspecified, entries refer to chimpanzees. Page references in **bold** represent figures/diagrams. Page references in *italics* represent whole chapters devoted to that topic.